CHEMICAL EXPOSURE
AND
TOXIC
RESPONSES

CHEMICAL
EXPOSURE
AND
TOXIC
RESPONSES

Edited by
Stephen K. Hall
Joana Chakraborty
Randall J. Ruch

LEWIS PUBLISHERS

Boca Raton   New York   London   Tokyo

Acquiring Editor:	Ken McCombs
Project Editor:	Suzanne Lassandro
Marketing Manager:	Greg Daurelle
Direct Marketing Manager:	Arline Massey
Cover design:	Denise Craig
PrePress:	Carlos Esser
Manufacturing:	Sheri Schwartz

Library of Congress Cataloging-in-Publication Data

Chemical exposure and toxic responses / editors, Stephen K. Hall, Joana Chakraborty,
Randall J. Ruch.
      p.   cm.
      Includes bibliographical references and index.
      ISBN 1-56670-239-9 (alk. paper)
      1. Industrial toxicology. 2. Toxicology. I. Hall, Stephen K. II. Chakraborty,
Joana. III. Ruch, Randall J.
      RA1229.C457 1996
      615.9'02—dc20   96-20459
                                                                 CIP

© 1997 by CRC Press, Inc.
Lewis Publishers is an imprint of CRC Press

No claim to original U.S. Government works
International Standard Book Number 1-56670-239-9
Library of Congress Card Number 96-20459
Printed in the United States of America 1 2 3 4 5 6 7 8 9 0
Printed on acid-free paper

Preface

Exposure to toxic chemicals and its biologic responses has become a major focus for an increasing number of health professionals. Industrial hygienists, safety professionals, public and environmental health specialists, nurses, physicians, emergency planners and responders, and many other professionals are all intimately involved in the evaluation and analysis of exposures to toxic chemicals. Over the past decade, the scope and importance of hazardous materials toxicology has expanded beyond health professionals, and now impacts an enormous number of nonhealth-related groups such as financial institutions, city, state, and county planning boards, private corporations, and many other organizations.

Chemical Exposure and Toxic Responses attempts to present the myriad actual and potential health implications of hazardous chemicals in a single source. This book is organized so that one can proceed from a general perspective on the scope of the problem of chemical exposure and toxic responses to an understanding of toxicology and a method of inquiry. It is imperative that the persons who make the decisions to manufacture and use toxic chemicals be aware of their responsibility for potential effects on workers, users of products, and the environment.

Chemical Exposure and Toxic Responses is the product of the authors' and editors' involvement in industrial toxicology. Written for those who need practical toxicological information, the book compactly and efficiently presents the scientific basis of toxicology as it applies to the workplace, covers the diverse chemical hazards encountered in the modern work environment, and provides a practical understanding of these hazards for those concerned with protecting the health and well being of people at work.

Chemical Exposure and Toxic Responses consists of three parts: Part I establishes the general principles of industrial toxicology; Part II addresses specific effects of toxic agents on specific physiological organs and systems; and Part III is devoted to the evaluation of hazards in the workplace.

Editor

Stephen K. Hall. With more than 20 years experience in occupational health and safety, Stephen K. Hall is a well-known corporate consultant on toxicology, industrial hygiene, and chemical safety. Dr. Hall has made a career of successfully helping clients achieve compliance with federal and state occupational safety and health regulations. He holds certifications as an industrial hygienist in toxicological aspects, a hazardous materials manager, a hazard control manager, a hazardous waste specialist, and a clinical laboratory scientist. He has served as dean of health sciences at several academic institutions and is the author of numerous books and journal articles on toxicology and other aspects of chemical safety.

Associate Editors

Joana Chakraborty. A scholar, scientist, and educator, Dr. Joana Chakraborty is a professor in the Department of Physiology and Molecular Medicine at the Medical College of Ohio at Toledo. For the past 26 years, she has been involved in teaching medical, nursing, and physical therapy students and made major contributions in the area of reproductive physiology. Dr. Chakraborty has published numerous scientific articles in peer-reviewed journals and presented papers in national and international meetings. She is a recipient of many prizes, including a post-doctoral fellowship from the National Institute of Health and a fellowship from the Ford Foundation.

Randall J. Ruch. Dr. Randall J. Ruch has been an assistant professor of toxicology in the Department of Pathology at the Medical College of Ohio since 1992. Currently, he is the course director of Cellular and Molecular Toxicology and participates in teaching other toxicology, oncology, and pathology courses at the Medical College of Ohio. His research in experimental carcinogenesis and hepatotoxicity is funded by the National Cancer Institute, The Department of Defense, and the American Institute for Cancer Research.

Contributors

Joana Chakraborty
Medical College of Ohio
Toledo, Ohio

Cathleen M. Crawford
Crawford Communication
  Consultants
Palos Heights, Illinois

Robert B. Forney, Jr.
Medical College of Ohio
Toledo, Ohio

Amira F. Gohara
Medical College of Ohio
Toledo, Ohio

Peter J. Goldblatt
Medical College of Ohio
Toledo, Ohio

Stephen K. Hall
Envirotox, Inc.
San Francisco, California

James A. Hampton
Medical College of Ohio
Toledo, Ohio

Maureen M. McCorquodale
University of Illinois at Chicago
Chicago, Illinois

Randall J. Ruch
Medical College of Ohio
Toledo, Ohio

Herman A. J. Schut
Medical College of Ohio
Toledo, Ohio

Joseph C. Siglin
Medical College of Ohio
Toledo, Ohio

Gary D. Stoner
The Ohio State University
Columbus, Ohio

Table of Contents

Part I: General Principles

Part II: Toxic Responses

Part III: The Work Environment

Part I: General Principles

1

Basic Concepts of Exposure and Response

Robert B. Forney, Jr.

CONTENTS

INTRODUCTION

The study of the interactions which occur between a toxicant and an individual may be divided into two parts. The mechanisms of injury and the signs and symptoms associated with it are the "dynamics" of the toxicant's action. However, before such dynamics may occur, a toxicant must first accumulate at a target organ or cell in adequate concentration to exert its effect. The study of the transit of a toxicant into and through the body, of its metabolism and excretion, and of the rates at which these events occur is called "toxicokinetics".

Observed variability in response to toxic agents can be attributed to either dynamic or kinetic factors. Dynamic differences in response are due to the dose; host factors such as age, genetic background, and disease state; and environmental factors such as the form of the agent, atmospheric conditions, and the presence of other agents. Kinetic differences can be attributed to the route and chronicity of exposure, and the rates and extent of absorption, distribution, biotransformation, and excretion. These latter differences affect the "dose" at a target cell, and therefore, the outcome of the exposure. Although toxic effects are often listed for agents without reference to dose, it is the dose and its individual-specific kinetics which make the difference.

Until recently, it was believed that the skin, oropharynx, and respiratory system effectively prevented the entry of toxic agents into the body. Now it is known that toxicants can pass through these barriers, even though they may do so at a very slow rate. Generally speaking, inhalation is the most rapid, and dermal is the least rapid, route of entry. But the rate of entry is determined by the form of the agent, its dose, and the integrity and saturability of the epithelium involved.

In many cases, the route of absorption will determine not only the rate of accumulation but also the metabolism and effects of an agent. Ingestion exposes the gastrointestinal (GI) tract to direct contact with the parent toxicant. It also exposes the agent to the acid, pepsin, and bacterial flora of the GI tract. Unlike other routes of entry, all of the blood supply from the GI tract must transit the liver before entering the general circulation. As a result, much more of the agent is exposed to metabolic alteration before circulating to target organs.

The major routes of entry are the dermal, gastrointestinal, and respiratory. The relative importance of a given route will depend upon the chemical form of the agent, and the individual and environmental conditions (ventilation, protective barriers, other agents, etc.) present during the exposure.

Once an agent has entered the body, it is distributed to target sites of action — sites of metabolic change and excretion. The circulatory system is of prime importance, because it transports both water and lipid soluble agents via the plasma and the proteins contained within the plasma. The proteins bind some toxicants for release at different points within the system. Changes in binding, as well as changes in the target points for storage, action, metabolism, or elimination, will alter the toxic effects which result from the individual having been exposed.

Finally, the chronicity of the exposure must be considered. Acute (single or short duration exposure), subacute (several weeks), and chronic (months or years) each present further complicating factors which need to be considered before an exposure outcome can be accurately predicted.

DOSE-RESPONSE RELATIONSHIPS

The relationship between the dose of a toxicant and the resulting effect is the most fundamental aspect of toxicology. Many believe, incorrectly, that some

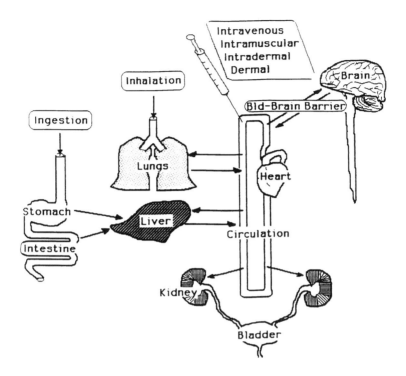

FIGURE 1.1 Adsorption and distribution of chemicals in the body.

agents are toxic and others are harmless. In fact, determinations of safety and hazard must always be related to dose. This includes a consideration of the form of the toxicant, the route of exposure, and the chronicity of exposure.

A dose-response relationship can be constructed for each type of effect an agent is capable of producing. These curves can be plotted for populations (the Y axis expressing epidemiological data in percentages) or for individuals (the Y axis expressing a quantitative measure of response, e.g., beats per minute for change in heart rate).

Some terms frequently used in describing the dose-response relationship of a given toxicant are given in Table 1.1.

CHRONICITY OF EXPOSURE

There are four classifications of toxicant exposure by duration or chronicity:

Classification	Duration of Exposure
Acute	Less than 24 h
Subacute	1–30 days
Subchronic	1–3 months
Chronic	More than 3 months

TABLE 1.1   Terms Commonly Used To Describe the Dose-Response Relationship of Toxicants

Potency	The range of *doses* over which an agent produces increasing effects. A more potent agent produces the same effect at a lower dose.
Efficacy	The range of *effects* which an agent is capable of producing when the doses is adequate. A more efficacious agent is capable of producing a magnitude of responses which is *not possible* for a less efficacious agent to produce.
Threshold	The minimally effect dose of an agent necessary to produce a given effect.
Hypersusceptible	Individuals who respond at doses *significantly lower* than the mean for their population.
Resistant	Individuals who respond at doses *significantly higher* than the mean for their population.
NED	Normal equivalent deviations. An expression of variance in a normally distributed population responding to increasing doses of a toxicant. NED's range typically from -3 to +3 with 0 NED representing the population mean.
Probit	NED + 5. Suggested by Bliss in 1935 as a method to eliminate negative values in plotting NED data. Probits therefore typically range from 2 to 8, with 5 representing the population mean.

% Response	NED	Probit
0.1	-3	2
2.3	-2	3
15.9	-1	4
50.0	0	5
84.1	+1	6
97.7	+2	7
99.9	+3	8

LD_{50}	Median Lethal Dose, lethal to 50% of a population.
ED_{50}	Median Effective Dose, effective to 50% of a population.
Therapeutic Index	LD_{50}/ED_{50} An attempt to measure the "safety" of a compound. Used previously for medications.
Margin of Safety	LD_1/ED_{99} This improved expression is used primarily for single dose comparisons.
NOEL or NEL	No Observed Effect Level or No Effect Level.
Risk Extrapolation	Process of evaluating risks associated with an exposure to a given agent.

The duration of exposure may be the most important determinant of the dose, and therefore, of the agent's toxicity. Acute exposure often is by a single dose, but repeated exposures still qualify as long as they occur within a single 24-h period. Repeated doses will lead to accumulation if the interval of exposure is less than the time required for elimination. The rate of accumulation is increased as the exposure interval relative to the elimination rate is decreased.

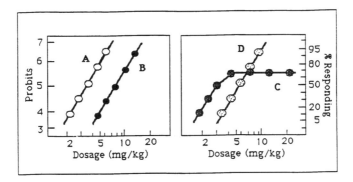

FIGURE 1.2 Illustrations of the concepts of "potency" ("A" is more potent than "B"), "efficacy" ("D" is more efficacious than "C") and "probits" (compare the probit scale on the left with the % response scale on the right.)

Occupational exposures are often chronic, with "exposure holidays" occurring when the worker is away from the workplace. An exposure may be chronic without representing a hazard to the individual if the dose is low enough, or if the interval of exposure is long enough. Improvements in the use of protective barriers and monitoring devices have generally reduced hazards.

Airborne toxicants may have reference ranges established for an index of their hazard. "Threshold Limit Values" (TLVs, published by the American Conference of Governmental and Industrial Hygienists) take chronicity into account. Three chronicity ranges are used in combination:

TLV	Meaning	Duration Covered
TWA	Time Weighted Average	Average concentration safe for 8-h working day
STEL	Short Term Exposure Limit	Average concentration safe for 15-min period
Ceiling	Maximum safe concentration	Single moment concentration

TYPES OF TOXIC REACTIONS

Not all of the effects produced by toxicants are harmful. The harmful ones may be described by one or more types of reactions and are given in Table 1.2.

STRUCTURAL CONSIDERATIONS

Membranes

Davson and Danielli first postulated the basic bimolecular lipid leaflet structure of cell membranes. With few exceptions (occurring only in some specialized plant and animal species), this basic nature of membranes is consistent. The epithelial cells lining the GI and respiratory tracts have membranes of this nature.

TABLE 1.2   Types of Toxic Reactions

Reaction Type	Description
Allergic	Requires previous sensitization by the agent or by another substance structurally similar to the agent
Idiosyncratic	Genetically determined *abnormal* response
Local	Response at the site or sites of contact with the agent
Systemic	Response following absorption and distribution of agent
Immediate	Response develops rapidly after single exposure
Delayed	Response which develops after a "latent" period (e.g., cancer)
Reversible	Tissue has capacity for regeneration from injury
Irreversible	Tissue injury is such that regeneration is not possible
Additive	Two or more agents exerting combined effect which is *equal* to the sum of each individual effect
Synergistic	Two or more agents exerting combined effect which is *much greater than* the sum of each individual effect
Antagonistic	Two or more agents exerting combined effect which is *much less than* the sum of each individual effect; this may be due to offsetting effects, blocking effects, inactivation or change in absorption, distribution or elimination
Tolerance	State of diminished effectiveness of a toxicant resulting from prior exposure; tolerance may be due to either dynamic or kinetic alterations in responsiveness

Two mirror image, lipid leaflets form opposing sides of the membrane, giving it an average width of 75 Å. Phospholipids and cholesterol are the predominant lipids found with spingolipids and other minor components. For this reason, lipid-soluble toxicant forms often pass more freely through the membrane than water-soluble ones.

Glycoproteins are distributed throughout the lipid bilayer. Although some of these proteins move freely within, others stabilize it by serving as anchor points for microfilaments.

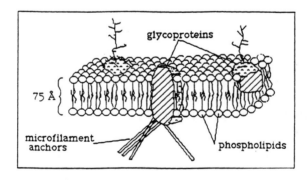

FIGURE 1.3   Bimolecular, lipid leaflet membrane model.

The ratio of lipid to protein varies from 5:1 for myelin membranes in the nervous system, to 1:5 for the inner structure of mitochondria. The membrane surface also varies in the ratio of lipids to proteins. The extent and rate of absorption may be more dependent upon this ratio than on the protein content of the membrane as a whole.

BIOCHEMICAL CONSIDERATIONS

Membrane permeability is influenced by many toxicant-specific, biochemical factors. Two of the most important are the degree to which an agent is ionized, and its inherent lipid solubility as measured by its "partition coefficient". These factors are interrelated. The lipid solubility is affected by the degree of ionization, with non-ionized species being more lipid soluble and, therefore, more likely to pass through membranes quickly.

Ionization

The effect of ionization on absorption was first demonstrated for strychnine. In stomach acid, this alkaline toxicant is highly ionized, while in the alkaline environment of the small intestine, it is non-ionized and easily absorbed. When the intestine was ligated at the stomach in an experiment, the toxic effects of strychnine were not observed. Ionization depends upon the toxicant's properties and the nature of the medium in which it is found.

Most agents are not capable of ionization. The alkaloids and organic acids are the most important toxicants affected by alterations in the pH of the medium in which they are contained. Alkaloids are less ionized in alkaline solutions, and organic acids are less ionized in acidic solutions.

The ionization of a given agent is described by its "pK_a". The pK_a is equal to the pH of a solution at which 50% ionization is observed. The Henderson-Hasselbach equation gives these relationships:

$$\text{For acids:} \quad pK_a - pH = \log \text{[non-ionized]/[ionized]} \tag{1}$$

$$\text{For bases:} \quad pK_a - pH = \log \text{[ionized]/[non-ionized]} \tag{2}$$

When an agent is ionized, it reacts with lipids and proteins in a way that limits membrane transversal. On the other hand, almost all agents will cross membranes to some extent, regardless of their degree of ionization. Some ionized agents are readily absorbed. The herbicide paraquat and the organophosphate antidote, pralidoxime (2-PAM) are good examples of this.

Partition Coefficient

The partition coefficient is a measure of an agent's distribution between an aqueous and lipid phase of a solution. It is an attempt to quantify, for comparison purposes, an agent's lipid solubility. It is given by the expression:

$$\text{Partition Coefficient} = \frac{\text{concentration in lipid phase}}{\text{concentration in aqueous phase}} \quad (3)$$

A high partition coefficient means that a toxicant is lipophilic. A good correlation has been demonstrated for the partitioning of the alcohols and their speed of absorption. But this correlation has not been demonstrated for other types of agents. A very highly lipid-soluble agent may penetrate a membrane well but not exit. The size of the molecule is another important factor, too. Generally speaking, large water-soluble molecules (lipid insoluble) are poorly absorbed.

ABSORPTION

Mechanisms

Although not well studied, toxicants cross membranes by four different mechanisms. The most predominant one is "passive diffusion". The agent must be lipid soluble to some degree and must be nonionized. The membranes of absorptive surfaces contain pores which permit small molecules to pass by "filtration". An agent must have a molecular weight of less than 100 Da to be absorbed in this way. Some toxicants are transported across membranes. Proteins "carry" them either passively, (requiring a concentration gradient but no energy) or actively (requiring energy but making it possible for the molecule to move against a concentration gradient). These mechanisms are called "facilitated transport" and "active transport", respectively. Finally, the membrane itself may be the mechanism of absorption. The membrane, in these cases, flows around an agent, allowing it to readily cross. If the agent is a liquid, this is called, "pinocytosis"; if it is a solid, it is termed, "phagocytosis".

In all cases in which absorption is passive, Fick's Law of diffusion will describe the determinants of the rate. They include the thickness of the membrane, the surface area available for absorption, and the concentration of the toxicant on both sides of the absorption surface. Since the portal bloodstream quickly carries absorbed material away from the GI tract, the concentration will approach zero on the inner surface, by that route.

The passive diffusion process is "first order", which means that it is dependent upon the concentration gradient that exists. The greater the gradient, the faster absorption will occur. An important factor to consider, however, is that the toxicant must be "in solution", before it will diffuse. Often, when drug overdoses are taken, the starch "excipients" (ingredients other than active drug placed in a pill or capsule) will not find adequate moisture to dissolve. This leaves a pill mass that delays absorption.

Charcoal is often used to bind toxicants in the GI tract and delay their absorption. For this strategy to work, much charcoal must be used (10 times the amount of agent ingested), and the agent must be non-ionized.

Another therapy is to reduce the concentration gradient by lavage or by stimulating emesis with ipecac. Both can be very effective at reducing the rate of absorption when they are performed quickly (less than 60 min post-ingestion) and properly.

Routes

The primary routes of entry into the body are dermal, gastrointestinal, and respiratory. Therapeutic agents and abused drugs may also be administered by injection. This adds several routes: intradermal, subcutaneous, and intravenous. Animal experiments often employ one last route: intraperitoneal. The first three will be discussed here.

Dermal

The skin is a complex, multilayered tissue which comprises about 10% of total body weight. It is relatively impermeable to most ions and aqueous solutions. It is, on the other hand, permeable to many toxicants. Examples of poisoning through dermal exposure include organophosphate pesticide poisoning of agricultural workers, chlorophenol in domestic animals, and solvents in factory workers.

The skin is composed of three distinct layers: epidermis, dermis, and subcutaneous. The epidermis is the most important outer barrier. It varies in thickness between 0.1 and 0.8 mm. It is composed of basal and columnar cells, with the highly keratinized stratum corneum on the surface. The cells are formed at the basal layer and change as they migrate towards the surface. The stratum corneum is the primary barrier to the penetration of toxicants.

The dermis is highly vascular; any agent which makes its way through the epidermis to it will be readily absorbed. Cuts and bruises to the epidermis therefore increase the risk of absorption of agents coming in contact with these wounds. The circulation of blood in the dermis is controlled so that exercise, temperature, and emotional state affect it and the absorption of toxicants.

The subcutaneous layer is mostly lipid and serves as a cushion and shock absorber, an insulator, and an energy reserve. It plays a minor role, however, in absorption.

The skin contains several other structures including: hair follicles, sebaceous glands, eccrine and apocrine sweat glands, and nails. Although percutaneous absorption could occur through hair follicles or sweat ducts, most lipid-soluble agents are believed to move directly as well. The enormous surface area of the skin (18,000 cm^2) is the most important factor.

Active transport has not been demonstrated, so penetration occurs by simple passive diffusion.

Protective barriers must repel toxicants, not absorb them. Most clothing merely serves to hold toxicants against the skin. Irritants such as acids, alkalis, or hydrocarbons break down the stratum corneum, increasing absorption.

Gastrointestinal

The oral route is normally associated with accidental poisoning in children and with intentional ingestions (drug abuse and suicide) in adolescents and adults. But this route is also commonly involved in industrial exposure. Workers who smoke, drink, or eat in the workplace may unintentionally swallow toxicants which have contaminated their hands, cigarettes, or eating and drinking utensils.

Although the buccal cavity (nicotine) or the rectum (suppositories) occasionally are used to introduce drugs, the stomach and small intestine are the important sites in the gastrointestinal tract.

All of the blood which circulates through these organs must first pass through the liver before entering the systemic circulation. This anatomic relationship makes it possible for a significant portion of a toxicant to be metabolized and often detoxified before encountering targets of its toxic action. This relationship is protective in most cases, although some relatively inactive agents may be activated. When significant, either as a protective device or in activation, this phenomena is referred to as a "first-pass effect".

Some toxicants are secreted in the bile in the gall bladder. As the bile enters the small intestine, some agents may be altered, released and absorbed again. This cycle is termed "entero-hepatic circulation".

The pH of the stomach is acid and the intestine is slightly alkaline (as discussed above). Therefore, an agent will likely be more highly ionized in one location or the other. Some agents may change their form dramatically (hydrolysis) in one of these environments.

The surface area of the small intestine is much greater than the stomach. This is due not only to its great length but also to the presence of villi, finger-like structures shown in Figure 1.4.

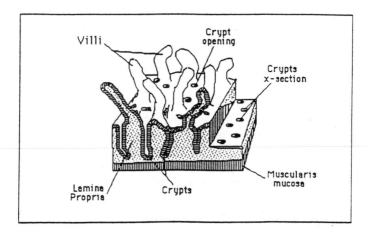

FIGURE 1.4   Small intestine lining showing villi and crypts.

Since most agents will be absorbed more quickly from the small intestine, the rate of stomach emptying is an important factor. All agents must dissolve before absorption can occur. Stomach emptying is generally delayed until the contents have been solubilized. The presence of food in the stomach will delay emptying. Excipients (nonactive ingredients in medications) are largely composed of starches. In drug overdoses, these materials often form concretions, sticky masses, which delay dissolution and therefore stomach emptying.

Other factors affecting absorption from the GI tract include particle size, solvent effect, the presence of emulsifiers, microorganisms normally present, binding to gut contents (e.g., charcoal, see above), intestinal motility (affected by many agents), and the health of the individual.

Respiratory

Without protection from filters or self contained breathing devices, the "inhalation" of airborne materials is unavoidable. Several anatomic features of the respiratory system are important in determining the rate and extent of absorption.

The respiratory tree presents a large surface area for absorption (100 m², 50 × the surface of the skin). Its moist lining traps water soluble agents near the oropharynx. The many changes in direction of the airway encourage particles to impact the lining in the upper portion of the tract. Cilia lining the airways stimulate the secretion of mucous to trap other agents and sweep the mucous to the throat to be swallowed and eliminated (a process called "broncho-toilet"). Irritant gases paralyze the sweeping of the cilia and stimulate mucous secretion. This results in accumulation of mucous in these exposures. "Phagocytes" engulf toxicants into lymph, where they may be stored for long periods. The airways gradually become smaller the deeper into the lung the air travels. "Particle size" and mass (ballistics) are therefore determinants of exposure severity and the location of injury. Particles 2 to 5 μm are deposited in the tracheobronchial region, while those smaller than 1 μm will generally reach the alveoli.

The blood circulates in very close proximity to the air in the alveolar sacs at the end of the respiratory tree. This allows the rapid absorption of agents able to make the trip, avoiding the protective design of the lung. The absorption is so quick that it mimics the intravenous route.

In spite of the large surface area of the lung, the constancy of the need for respiration (15 to 20 times per min), and the inevitability of exposure without protection, the lung is efficiently designed for self protection. This is illustrated by the fact that coal miners accumulate an average of only 100 g of coal dust (found at autopsy), even though they may inhale at least 6,000 g during their lifetimes.

There are several air volumes which are used to describe the process of respiration. The "dead space" is the volume of the respiratory tree apart from the alveolar region where oxygen and carbon dioxide are exchanged. The last 150 cc of air inspired and the first 150 cc exhaled are composed of "dead space" air. Emphysema results in a great increase in this volume and a decrease in the efficiency of respiration. The "vital capacity" is the volume of air involved in respiration when inspiration and expiration are performed with maximal effort. The "residual volume" is the volume of air remaining in the lung after a maximal expiratory effort. Gaseous agents are slowly released from the alveolar space on expiration due to the slow release of this volume. Therefore, several expirations are required to rid the lung of residual toxicant once an individual begins breathing clean air.

Irritant gases are classified by their water solubility. The more water soluble agents are removed in the upper respiratory tree:

Group	Water Solubility	Region	Examples
I	High	Nasopharynx	HCl, H_2SO_4, NH_4
II	Moderate	Tracheobronchia	Cl_2, NO_3
III	Low	Alveoli	$COCl_3$, SO_3

There is very little evidence of active transport occurring in the lung. The lung is, however, an important metabolic and endocrine organ.

DISTRIBUTION

After being absorbed, toxicants are distributed throughout the body. The rate of distribution is generally rapid, being determined by blood flow, lipid solubility (ability to cross membranes), protein binding, and the affinity a substance has for each tissue. Some toxicants pass membranes to a limited extent and have a restricted distribution. Others pass very well and are retained in the tissue to a great extent. After distribution, these substances will have very low plasma concentrations relative to the amount absorbed compared to those with more restricted distributions. The site of toxic action often is *not* the same as the site of principle storage.

Volume of Distribution

Body water can be divided into three "compartments". Plasma water is the smallest, representing 8% of the total, or about 3 L in an average 70 kg adult. Intrastitial water comprises 29%, or about 11 L. Taken together, plasma and intrastitial water are the extracellular fluid. Intracellular water is the largest volume, consisting of 24 L (63% of the total).

Highly ionized substances, like salicylate, do not leave the extracellular fluid to any appreciable extent. Salicylate's "volume of distribution" is approximately 10.5 L in a 70 kg adult (0.15 L/kg). However, distribution is complicated by protein binding and storage in tissues of great affinity to the particular agent. Lipophilic substances often "appear" to distribute in "volumes" much greater than the volume of the body, let alone of body water!

The apparent volume of distribution then, is a mathematical expression, useful to toxicologists, which does not necessarily correspond to any particular anatomic space or fluid compartment. After distribution is complete, there is a relationship between a dose absorbed and the plasma concentration which results. This relationship is given by the expression:

$$V_d = D_{iv}/C_{p0} \qquad (4)$$

where V_d is the apparent volume of distribution, D_{iv} is an intravenous dose administered (eliminating the problem of incomplete absorption), and C_{p0} is the plasma concentration extrapolated back to time zero. The plasma concentration will be incorrect at time zero unless back extrapolation is performed. After intravenous doses, the plasma concentration will be too high until distribution is complete (see Figure 1.5, graph on the right). Following oral administration, C_{p0} is too low (see Figure 1.5, graph on the left).

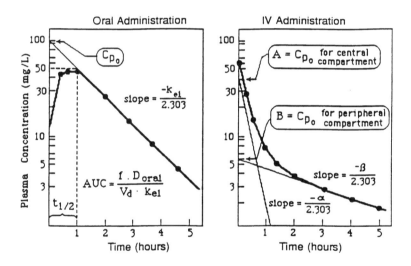

FIGURE 1.5 The graphic method of calculating kinetic parameters using a semilogarithmic plot of the plasma disappearance curve for an orally administered, one compartment model, agent (left); and, for an intravenously administered, two compartment model, agent (right).

Another expression for the volume of distribution is:

$$V_d = (f \cdot D_{iv})/(AUC_{0 \to \infty} \cdot k_{el}) \qquad (5)$$

where $AUC_{0 \to \infty}$ is the area under the plasma concentration versus time curve, from time zero to infinity; and k_{el} is the elimination rate constant. If the volume of distribution is calculated from equation 4, and $AUC_{0 \to \infty}$ can be estimated from repeated plasma concentration determinations, then k_{el} can be determined. If an oral administration (D_{oral}) is used, then the fraction, "f" of the dose which is absorbed (termed the "bioavailability") must be included.

$$V_d = (f \cdot D_{oral})/(AUC_{0\to\infty} \cdot k_{el}) \qquad (6)$$

Some agents are completely absorbed. In this case, "f" would be equal to 1.0. Often however, absorption may be incomplete. Failure to include this fact in the estimation leads to an inaccurate estimate of distribution. Bioavailability itself can be estimated by comparing oral and intravenous administrations.

$$\text{Bioavailability} = f = AUC_{0\to\infty}, oral/AUC_{0\to\infty}, iv \times 100 \qquad (7)$$

Storage

If a toxicant distributes rapidly into tissues, then the whole body appears to be a single "compartment". Many agents however, require more time for their concentrations in tissues to equilibrate with the plasma. In these cases, there appears to be two or more "compartments". As in the case of the volume of distribution, compartments are not anatomic locations. Rather, compartments are a mathematical way of describing the distribution.

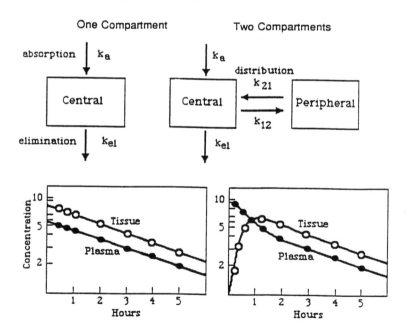

FIGURE 1.6 One and two compartment, open models of the absorption, distribution, and elimination of toxicants.

Some toxicants concentrate at the site of their greatest toxic effect. Carbon monoxide has a great affinity for hemoglobin, which is also the site of its toxicity. Some toxicants are kept from their sites of greatest effect by their affinity for other tissues. Lead is stored in the bone but affects the soft tissues. In fact, the diagnosis of lead poisoning may be missed, the symptoms having abated, if lead storage is not considered.

The principle storage tissues are listed in Table 1.4:

TABLE 1.4   Principle Storage Tissues

Storage Tissue	Types of agents
Plasma proteins	Many substances
albumin	many drugs, dyes, acid drugs Ca^{++}, Cu^{++}, Zn^{++}
transferrin	Fe^{++}
ceruloplasmin	Cu^{++}
α, β lipoproteins	vitamins, cholesterol, steroids, lipid soluble agents, basic drugs
gamma globulins	antigens (larger molecules)
Liver	Many substances
Y proteins (ligandin)	organic acids, azodye carcinogens, corticosteroids
metallothionein	Cadmium
Kidney	Many substances
metallothionein	Cadmium
Bone	Lead, Fluoride, Strontium
Fat	Lipophilic substances DDT, PCBs, PBBs, chlordane

Blood-Brain Barrier

Substances distribute to the central nervous system (CNS) less effectively than to other tissues given the extent of the blood supply which the brain receives. There are three anatomic and physiologic reasons for this (Figure 1.7):

1. The capillary endothelial cells of the CNS are tightly joined, eliminating the pores which occur in the membranes of other tissues.
2. Glial cell processes (astrocytes) surround the capillaries.
3. The protein concentration of the interstitial fluid of the CNS is much lower than in other tissues.

In contrast to other tissues, toxicants entering the CNS must overcome these restrictions. But not all parts of the CNS are equally protected by this "barrier". The cortex and parts of the hypothalamus and hypophysis are more permeable than other regions.

The blood-brain barrier is not completely developed at birth, so infants are generally more susceptible to CNS toxicants.

Placental Barrier

It was once thought that the placenta protected the fetus well from exposure to toxicants. Anatomically there are great interspecies differences in the composition of the barrier between maternal blood and the fetus. Therefore, con-

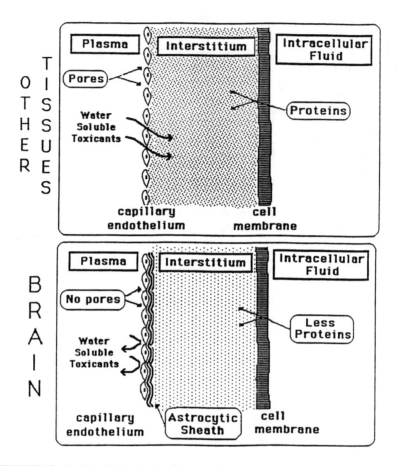

FIGURE 1.7   The blood-brain barrier of the central nervous system.

clusions drawn to man from studies performed in animals are tenuous. The human placenta has three cell layers and is called "hemochorial" because the chorionic villi of the placenta are in direct association with maternal blood. In domestic animals, there are as many as three additional cell layers separating the fetal chorion from material blood. It is reasonable to believe that the more cell layers there are, the less easily toxicants will pass into the fetus.

In all species, toxicants cross the "placental barrier" by passive diffusion. Although there are no identifiable "barrier" mechanisms beyond the number of cell layers (as in the CNS), some toxicants are detoxified by metabolism as they cross.

ELIMINATION

There are many different routes of elimination. More toxicants are eliminated by the kidney than by any other tissues. But the substance must be water soluble before it can be excreted in the urine. Many substances must be biotransformed to more water soluble forms first. Some specific agents will be principally excreted by routes other than the kidney. DDT and lead are eliminated by the liver and biliary system. Carbon monoxide is exhaled through the lung.

Elimination is a first order process. This means that the rate of elimination is dependent upon the concentration of the toxicant being eliminated. Higher concentrations of an agent are eliminated at a faster rate than lower concentrations. A semilogarithmic plot, depicted by Figure 1.5, produces a straight line after absorption and distribution are completed. The elimination rate (k_{el}) can be calculated from the slope of the elimination phase plotted in this way:

$$\text{slope} = -k_{el}/2.303 \qquad (8)$$

The time required for the concentration to decrease by one-half remains constant even though less is being eliminated. This time is called the half-life or $t_{1/2}$. It can be estimated graphically (Figure 1.6) or calculated simply from k_{el}:

$$K_{el} = 0.693/t_{1/2} \qquad (9)$$

Another mathematical term useful in describing a toxicant's elimination is the clearance, Cl. The total body clearance for a toxicant whose kinetics are best described by a one compartment open model is given by:

$$Cl = K_{el} \cdot V_d \qquad (10)$$

or, a model independent expression (valid for both one and two compartment model agents) is:

$$Cl = D_{iv}/AUC_{0 \to \infty} \qquad (11)$$

When the dose and the dosing interval are kept constant, the toxicant accumulates until the concentration achieved is great enough, for an amount equal to the dose to be eliminated during the dosing interval. At this point, the accumulation stops and an equilibrium condition called "steady-state" exists. This phenomenon is illustrated in Figure 1.8.

The average steady-state concentration, $\bar{C}_{ss}$ can be calculated when other kinetic parameters are known:

$$\bar{C}_{ss} = (f \cdot D_{oral})/(V_d \cdot k_{el} \cdot \gamma) \qquad (12)$$

The dosing interval is γ. Using equation #8, the calculation can be expressed in terms of $t_{1/2}$ by:

$$\bar{C}_{ss} = (1.44 \cdot t_{1/2} \cdot f \cdot D_{oral})/(V_d \cdot \gamma) \qquad (13)$$

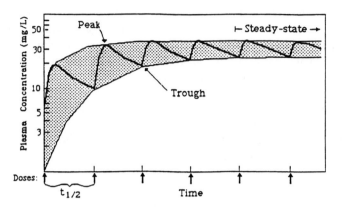

FIGURE 1.8   The accumulation of a toxicant in plasma which is administered repeated at a constant interval.

And finally, the average "body burden" of a toxicant, $\overline{X}_{ss}$, can be expressed:

$$\overline{X}_{ss} = \overline{C}_{ss} \cdot V_d \tag{14}$$

Exposure to toxicants occurs involuntarily through food, drink and air, and intentionally in the workplace, in therapy or in misuse of drugs or chemicals. Toxicants are absorbed, distributed, and eliminated by mechanisms and rates which vary depending upon the toxicant's physicochemical properties, the individual, and the environment. When exposure is repeated before elimination is complete, accumulation occurs. In this way, toxicity may result after an asymptomatic period, even though each dose was, by itself, below the threshold of toxicity.

REFERENCES

Hayes, A. W. (ed), *Principles and Methods of Toxicology*, 3rd edition, Raven Press, New York, 1994.

Hodgson, E. and Levi, P. E., *Introduction to Biochemical Toxicology*, 2nd edition, Appleton & Lange, Norwalk, CT, 1994.

Klaassen, C. D. (ed), *Cassarett and Doull's Toxicology: The Basic Science of Poisons*, 5th edition, Pergamon Press, New York, 1995.

Lu, F. C., *Basic Toxicology: Fundamentals, Target Organs, and Risk Assessment*, 2nd edition, Taylor & Francis, Bristol, PA, 1991.

Timbrell, J. A., *Principles of Biochemical Toxicology*, 2nd edition, Taylor & Francis, London, 1991.

2

Metal Exposure and Toxic Responses

Stephen K. Hall

CONTENTS

INTRODUCTION

The increasing technologic use of metals is one measure of progress of civi-
lization since early man's emergence from the Stone Age. This has posed

hazards, from the time metals were fashioned into spears to present-day exposures to space-age alloys. Today, metals are used extensively in the workplace and employee exposure can result from numerous industrial operations. Of the 80 or so elements that are classified as metals, more than half are of industrial and economic significance. As a group, metals exhibit a very wide range of biological, chemical, physical, and toxicological properties.

While a few metals have found some use in medicine, (e.g., inert metals such as tantalum, platinum, gold used as surgical implants, and lithium in the treatment of manic depression) all metals are potentially toxic at high concentrations, even though some of them are also essential. An essential metal is one for which a deficiency results in impairment of function that is relieved only by administration of that metal. At excess concentrations deleterious effects begin to set in, the biological response becomes unfavorable, and the substance becomes toxic. Some of these essential metals and their daily requirements are listed in Table 2.1.

TABLE 2.1   Essential Metals and Daily Requirements

Metal	Daily Requirement (mg)
Calcium	0.8
Chromium	0.1
Cobalt	3
Copper	3
Iron	10–20
Magnesium	700
Manganese	4
Molybdenum	0.3
Potassium	2,000–5,000
Selenium	100
Sodium	1,000–2,000
Zinc	15

Animals maintain a concentration of essential nutrients within the optimal range by a complex set of physiological reactions termed homeostasis. Concentrations of all essential metal ions are under homeostatic control. Toxicity is usually unrelated to essentiality. Almost any substance in excess ultimately becomes toxic.

METABOLISM OF METALS

Metabolism of metals refers to all of the processes by which metals are handled by the body. Absorption, storage, and elimination are the most important processes in the metabolism of inorganic metal compounds. Organometallic compounds are metabolized by different biochemical pathways.

Common routes of metal absorption are through inhalation and the gastrointestinal tract. Metals in the air may be inhaled as a vapor or as an aerosol (i.e., fumes or dust particulate). Fumes and vapors of cadmium and mercury are readily absorbed from the alveolar space, as are many organometallic compounds such as tetraethyl lead. Large particles of metal aerosol (>10 μm) are trapped by the upper respiratory tract. They are then cleared by the muco-ciliary escalator, swallowed, and absorbed into the blood through the gas-trointestinal tract. Small particles (< 5 μm) may reach the alveolar space or gas exchange portion of the lungs. If they are water soluble, they are then rapidly absorbed from the alveoli into the blood. Organometallic compounds may also be absorbed through the skin.

Gastrointestinal absorption of metals and their compounds varies widely and depends on many factors such as the solubility, the chemical form, the presence of other materials, the competition for binding sites, and the physi-ological state of the gastrointestinal tract. For example, the salts of cadmium, lead, and tin are poorly absorbed (<10%), while the salts of arsenic and thallium are almost completely absorbed (>90%).

After metal ions enter the plasma, they are available for distribution throughout the body. Distribution usually occurs rapidly and the rate of dis-tribution to the tissues of each organ is determined by the blood flow through the organ. The eventual distribution of a metal compound is largely dependent on the ability of the chemical to pass through the cell membrane and its affinity for the various binding sites.

Toxicants are often concentrated in a specific tissue. The compartment where the toxicant is concentrated can be thought of as a storage depot (e.g., lead is stored in bone). While stored, the toxicant seldom harms the hosts. Storage depots, therefore, could be considered as protective mechanisms, preventing the accumulation of high concentrations of toxicants at the site of toxic action. The toxicants in these depots are always in equilibrium with free toxicant in plasma, and as the chemical is excreted from the body, more is released from the storage site.

Most metal ions have a high affinity for functionally essential amino acid side chains such as sulfhydryl, histidyl, or carboxyl groups, and can react directly with proteins to alter enzymatic function. They may also bind to cofactors, vitamins, and substrates, thereby altering the availability of these cell constituents for biological function. Since particular metals play essential roles in enzyme catalysis, such biological processes are sensitive to alteration by toxic metals that are chemically similar to the essential metal. Finally, the presence of an excess of any metal may lead to depletion of essential metals.

CHELATION THERAPY

Chelation is the formation of a metal ion complex in which the metal ion is associated with one or more electron donors referred to as a ligand. The ligand may be monodentate, bidentate, or multidentate (i.e., it may attach or coordi-nate using one or two or more donor atoms). Bidentate ligands form ring structures that include the metal ion and the two ligand atoms that are attached to the metal. The donor molecule is properly referred to as a chelating agent. This term is derived from the Greek *chela*, for claw.

Chelating agents are generally nonspecific in regard to their affinity for metals. To varying degrees they will mobilize and enhance the excretion of a wide range of metals, including essential metals in the body. Therefore, the use of chelating agents as a prophylactic without close medical supervision is condemned by the medical profession. Among the common chelating agents in use today are diethylenetriaminepentaacetic acid and dimercaptosuccinic acid.

TOXIC EFFECTS OF SELECTED METALS

Following are summaries of toxic effects of metals. These metals were selected for discussion because of their economic value and industrial use, and, more importantly, because their toxic effects have been documented. Each summary contains the clinical manifestation of signs and symptoms of toxicity, target organs or systems, metabolism, carcinogenicity, and detoxification. Information on absorption, distribution, excretion, and biological monitoring will be covered in other chapters.

Aluminum

Aluminum is presently considered a nonessential element in humans. A variety of biochemical interactions with aluminum have been investigated. Many of these effects pertain to aluminum neurotoxicity. Acute neurological disorders termed "dialysis dementia" or "dialysis encephalopathy" have been reported in chronic hemodialysis patients. Patients developing such neurological disorders have been shown to have high levels of aluminum in their serum and brain tissue. Perhaps the most intriguing and potentially significant aspect of aluminum toxicity is the evidence implicating aluminum in the pathogenesis of Alzheimer's disease, an insidious neurodegenerative disorder of unknown etiology for which there is no effective treatment or cure.

Pulmonary effects of aluminum occur following inhalation of bauxite fumes. The resulting pulmonary fibrosis producing both restrictive and obstructive pulmonary function impairment has been described by Schaver. Interestingly, inhalation of aluminum mists was used during the 1930s as a prophylaxis for pulmonary fibrosis resulting from inhalation of silica particles.

Aluminum compounds can affect absorption of other elements in the gastrointestinal tract and alter intestinal function. Aluminum inhibits fluoride absorption and may decrease the absorption of calcium and iron. The binding of phosphorus or the interference with phosphate absorption can lead to osteomalacia, a condition marked by softening of the bones.

Absorbed aluminum is highly protein-bound, making it difficult to remove. In addition, there are large stores in bone and nervous tissue that are difficult to access. Chelation of aluminum using deferrioxamine alters the dialyzability of aluminum and facilitates its removal by hemodialysis.

Antimony

Antimony and its compounds are not essential to humans. Antimony poisonings today are rare in highly developed countries because of limited commercial and industrial use. The toxic effects of antimony compounds are similar to those of the corresponding arsenic compounds. There are two oxidation states for antimony: the trivalent form and the pentavalent form. The trivalent forms are more toxic than pentavalent forms, but only a small fraction of the pentavalent is reduced to the trivalent form in the liver. The disposition of the trivalent and pentavalent forms differs. Trivalent antimony is concentrated in red blood cells and liver, whereas the pentavalent form is mostly in plasma. Both forms are excreted in feces and urine; more pentavalent antimony is excreted in urine, whereas more trivalent antimony is excreted gastrointestinally.

Chronic effects due to antimony and its compounds are alterations of the electrocardiogram, especially T-wave abnormalities, and myocardial changes. Other chronic effects include liver toxicity, characterized by jaundice and fatty degeneration, pulmonary congestion and edema, and pustular skin eruptions. Antimony miners have developed disabling, but benign forms of pneumoconiosis.

The most toxic antimony compounds are stibine (SbH_3), the tri- and pentachloride, and the tri- and pentafluoride. Stibine is a powerful hemolytic poison. One of the earliest signs of overexposure is hemoglobinuria. Fatalities caused by antimony pentachloride have been reported. After oral ingestion of toxic antimonial compounds, irrigating of the stomach, administration of 2,3-dimercaptopropanol, and a liver-protecting therapy are recommended.

Arsenic

Arsenic is one of the oldest poisons used by man. The effects of arsenic poisoning were described in detail in the pre-Christian era. In general, the naturally occurring arsenic is pentavalent, while that added to the environment is trivalent. Trivalent arsenic compounds are more toxic than pentavalent compounds, and natural oxidation favors the conversion of trivalent arsenic to the pentavalent form.

There are two forms in acute arsenic poisoning: the paralytic form and the gastrointestinal form. The paralytic form is observed if large doses of arsenic are absorbed quickly. Symptoms develop quickly and are usually characterized by constriction of the throat followed by dysphagia. Death may occur by general paralysis. In the gastrointestinal form, abnormal symptoms such as watery diarrhea caused by paralysis of the central mechanism of the capillary control in the intestinal tract are dominant. This may result in a decrease in blood volume and blood pressure falls to shock levels. This in turn results in disturbed heart action and causes death.

Chronic arsenic poisoning may show many forms. The skin is the major organ of interest since most effects are first seen here. It begins as local erythema with burning and itching, giving the skin a mottled appearance. This is followed by swelling and sometimes vesicular eruptions. Melanosis is first seen on the eyelids, temples, and neck, and then spreads through the trunk. This is also often accompanied by hyperkeratosis, hyperhidrosis (excessive sweating on palms and soles), and warts. Mee's lines (horizontal white bands across nails) are also commonly seen and are considered to be a diagnostic accompaniment of chronic arsenic poisoning. Nasal septum ulceration is seen after long industrial exposure.

Arsine (AsH_3) is the most toxic arsenic compound. It is a colorless gas with a slight garlic-like odor and can be generated by side reactions or unexpectedly. Once absorbed, arsine is gradually oxidized in the body to arsenic trioxide. During this oxidation the protein of the red blood cells is denatured, resulting in hemolysis. Invariably, the first sign observed is hemoglobinuria, coloring the urine a port wine hue. Jaundice starts at the second or third day and rapidly spreads over the whole body. As a result of the rapid destruction of the red blood cells, large quantities of free hemoglobin block the renal tubules with hemoglobin crystals and fragments of cells. This is manifested by increasing oliguria, followed by anuria, leading to uremia and death.

While there has been some controversy, the epidemiologic evidence indicates that industrial and agricultural exposure to arsenic is implicated in cancers of the skin and respiratory tract. The individuals at greatest risk are smelter workers. For this reason, the International Agency for Research on Cancer (IARC) has declared that "there is sufficient evidence that inorganic arsenic compounds are skin and lung carcinogens in humans." It is interesting to note that up to now it has only been possible to show that arsenic has a mutagenic as well as a teratogenic effect in animal experiments; and no laboratories have successfully produced malignant tumors in animals.

In Britain, during the Second World War, the so-called British Anti-Lewisite (BAL), 2,3-dimercaptopropanol, a bidentate chelating agent, was developed in order to protect against an attack with arsenic-containing chemical warfare agents. The mechanism of action is to compete with enzymes in the reactions for arsenic-containing compounds since by the formation of a stable 5-membered ring, arsenic is strongly bound. Other dithiols such as sodium 2,3-dimercaptopropanesulfonate and 2,3-dimercaptosuccinic acid have also been found to be effective chelating agents. In arsine poisoning, however, the condition can only be treated symptomatically. Chelating agents are not effective.

Barium

Barium is not an essential element in the body. Barium ions are taken up, retained, and excreted in much the same way as calcium ions. While the insoluble forms of barium, particularly barium sulfate, are not toxic by the oral route because of minimal absorption, the soluble barium compounds are quite toxic. Accidental poisoning from ingestion of soluble barium compounds has resulted in gastroenteritis, muscular paralysis, decreased pulse rate, and ventricular fibrillation. The toxicity is related to a competitive inhibition of potassium ions and removal of sulfate ions. Baritosis, a benign pneumoconiosis, arises from the inhalation of barium sulfate dust and barium carbonate. It is not incapacitating, but does produce radiologic changes in the lungs. The radiologic changes are reversible with cessation of exposure.

In the rare case of acute barium poisoning, subcutaneous atropine injections to prevent colic and respiration-supporting measures are recommended. In the case of electrocardiogram changes or hypokalemia symptoms, a potassium infusion therapy should be of value. Chelating agents are not effective in barium poisoning.

Beryllium

Beryllium is not an essential element in the body. Beryllium and its compounds form insoluble precipitates at acid pH and are rarely absorbed from the gastrointestinal tract. However, contact with water-soluble beryllium salts may result in acute dermatitis and/or ocular lesions. Cutaneous injuries from beryllium metal, alloys, or oxide may require surgical treatment with excision of the foreign substance. Long-term skin contact of beryllium and its compounds may result in contact dermatitis and skin sensitization. In the bloodstream, a colloidal beryllium phosphate is formed and deposited in the liver, spleen, and bone marrow.

The most dangerous mode of entry into the body is by the respiratory tract, and effects are predominantly pulmonary. Pulmonary beryllium disease has two different forms: an acute chemical tracheobronchopneumonia, and a chronic form with granulomatous lesions of the lung. Acute pulmonary beryllium diseases are caused by the inhalation of large doses of beryllium or its soluble compounds in finely dispersed forms. The acute forms correlate with the concentration of beryllium exposure.

Chronic pulmonary beryllium disease, or berylliosis, is quite different from that of the acute form, with severe shortness of breath being the leading symptom. Pulmonary X-rays show miliary mottling, a haziness to "snow-flurry" effect. Histopathological examination of lung tissue shows interstitial granulomatosis. A dose-response relationship between extent of exposure and severity of disease is distinctly absent. As the fibrosis increases, bleb formation is common and pneumothorax occurs. This progressive form of berylliosis is accompanied by a decreased life expectancy.

Chelating agents for removal of deposited tissue beryllium have been explored. Among these, aurintricarboxylic acid has been found to be effective in mice and rats if given parenterally within eight hours after intravenous injection of an otherwise lethal dose of beryllium sulfate. However, chelating agents are not effective in chronic beryllium poisoning cases.

Cadmium

Cadmium is not an essential element in the body. The main organs for cadmium accumulation in humans are the liver and kidneys. After absorption, cadmium is transported to the liver where it stimulates synthesis of metallothionein.

Cadmium bound to metallothionein released from the liver moves via the blood to the kidney. The kidney is a critical organ in long-term, low-level exposure. In the kidney, cadmium-metallothionein accumulates in tubular cells. Since cadmium first affects the proximal tubule's reabsorption capabilities, the first effect to be detected is proteinuria. Later signs include amino aciduria, glucosuria, decreased urine-concentrating ability, and abnormalities in handling uric acid, calcium and phosphorus. The mineral problems may lead to kidney stones and osteomalacia (i.e., softening of the bone due to mineral loss).

Acute toxicity of cadmium in the work place is usually via the respiratory route and is characterized by lung edema, cell proliferation, and fibrosis. Symptoms of respiratory exposure in the work place include coughing, shortness of breath, irritation of the upper respiratory tract, and loss of sense of smell. Yellow staining of the teeth in heavy industrial exposure has been reported.

Itai-itai (ouch-ouch) disease was reported in Japan. This cadmium-induced disease was a result of dietary intake of the metal. Symptoms include painful sites spread all over the body, and difficulty in walking. The condition progresses, and bone fractures are common. Pathological changes include osteomalacia and osteoporosis, especially prevalent in postmenopausal women with deficient diets.

An increase in carcinoma of the prostate was first noted in a mortality study of battery workers in England but this was not found in a study of a large worker population. The National Institute for Occupational Safety and Health (NIOSH) has reviewed a number of epidemiological studies and recommended that cadmium and its compounds be regarded as potential occupational carcinogens.

Chelation therapy is not available for cadmium toxicity in humans. Experimental studies have shown that the action of chelating agents on the pharmacokinetics of cadmium depends on the time of administration of the agent after cadmium exposure. When chelating agents are given shortly after cadmium exposure and before new metallothionein has been synthesized, the dithiols such as 2,3-dimercaptopropanol and penicillamine increase the biliary excretion of cadmium while ethylenediaminetetraacetic acid, diethylenetriaminepentaacetic acid, and related chelating agents increase urinary excretion. For chronic exposure, when cadmium is bound to metallothionein, there is little effect from chelation therapy. In the case of itai-itai, large doses of vitamin D given over a period of months are effective in relieving painful symptoms.

Chromium

Chromium exists in several valence states. Only the trivalent and hexavalent are biologically significant. Trivalent chromium is an essential element in the body. It plays a role in glucose and lipid metabolism. Trivalent chromium is poorly absorbed by the body regardless of the route of administration, while hexavalent chromium is more readily absorbed. Hexavalent chromium can be reduced to trivalent by the gastrointestinal tract, thereby reducing uptake.

In occupational exposures, the lungs are the primary route; skin is considered a minor route of entry for both trivalent and hexavalent chromium compounds. In general, chromium compounds in the trivalent state are of a low order of toxicity. The toxic effects of occupational exposure to high levels of chromium have been recognized for more than 200 years. The overt signs of chromium toxicity (e.g., perforation of the nasal septum, skin ulcers or chrome holes, and liver and kidney damage) are rarely seen today. A common human toxicity associated with chromium exposure is allergic contact dermatitis.

Lung cancer is considered to be an occupational hazard for workers exposed to chromium in a wide variety of industrial and commercial occupations. The majority of information on chromium-induced carcinogenesis comes from human epidemiology studies of occupationally exposed workers. There is a good correlation between the dose of chromium and the relative risk of developing lung cancer. Hexavalent chromium compounds induce cancer in experimental animals at the site of exposure, while trivalent chromium compounds are inactive. However, there has been little success in inducing cancer in animals by topical application to the skin, and in humans there is no evidence for an increased risk of skin cancer, even at sites of severe skin ulceration.

The acute clinical manifestations of chromium toxicity are rarely seen today and there has been only limited investigation of systemic chromium detoxification.

Cobalt

Cobalt is essential as a component of vitamin B_{12} required for the population of red blood cells and prevention of pernicious anemia. The effect that cobalt enhances the formation of red blood cells and hemoglobin has been used for therapeutic purposes. This therapy, however, led to hypothyroidism and thyroid hyperplasia. Acute intoxications caused by repeated ingestion of soluble cobalt salts occurred in Canada and the United States. In these cases, cobalt salts had been added to beer as a foam stabilizer. Excessive beer drinkers showed cardiopathy. Up to 50% of these cases were fatal. Extreme cardiomegaly was accompanied by lower systolic blood pressure. Drinkers of cobalt-containing beer also suffered from hypothyroidism and thyroid hyperplasia. NIOSH has recommended thyroid palpation as part of the medical surveillance of workers exposed to cobalt.

Inhalation of cobalt dust may result in pulmonary fibrosis. This so-called hard-metal disease is a disease of the compliance of the lung caused by fibrotic alteration of the tissue. At the work place of hard-metal industry, cobalt occurs together with the carbides of tungsten or titanium. In spite of this, it is commonly believed today that cobalt itself is the etiologic factor of hard-metal disease. Occupational exposure to cobalt also gives rise to respiratory sensitization, bronchitis, asthma, and emphysema. Dependent upon the concentration of cobalt in the air, increases and decreases of red blood cell count and the content of hemoglobin have been observed.

Contact of cobalt with the skin may lead to dermatitis or eczema, especially if there are skin lesions. Allergic dermatitis of an erythematous papular type may also occur, and affected persons may show positive skin tests.

Little is known about therapy for acute or chronic human exposure to cobalt. It has been postulated that cobalt complexes are formed *in vitro*. The formation of these chelates might prevent cobalt from binding with biochemically important compounds containing sulfhydryl groups and thus might reduce the toxic response of reabsorbed cobalt ions.

Copper

Copper is an essential element for all vertebrates. It is part of several of the most vital enzymes, including tyrosinase, cytochrome oxidase, and amine oxidases. It is essential in the incorporation of iron into hemoglobin. The primary target organ for accumulation of copper is the liver, the metabolism of copper involves the turnover of the copper-containing enzymes.

There are two genetically inherited inborn errors of copper metabolism that are in a sense a form of copper toxicity. Wilson's disease is characterized by excessive accumulation of copper in liver, brain, kidneys, and cornea. Affected individuals fall largely into two broad categories: one with symptoms involving primarily the nervous system; the other involving the liver. When the accumulated copper is removed by effective chelation therapy, the symptoms of the disease regress. Wilson's disease is apparently synonymous with copper toxicity. Menke's disease, apparently an inability to absorb copper, is a copper deficiency. This disease is characterized by rapidly progressive cerebral degeneration and the presence of spirally twisted hair. It appears that in Menke's disease, copper is available but it is not metabolized in the normal manner.

Copper in dust, fumes, or sprays may produce "brass chills" which is a form of "metal fume fever." Chronic exposure may result in nasal ulceration and bleeding. Various damage to lung tissue has been reported in cases of severe intoxication after exposure to copper sulfate sprays. Exposure to copper dust may also cause discoloration of the skin. Although contact allergy to copper is very rare, there are reports of isolated incidences due to exposure to copper metals or salts. Oral ingestion may cause nausea, vomiting, diarrhea, and intestinal cramps. Liver and kidney functions may be disrupted giving rise to jaundice and cirrhosis. In severe cases of intoxication, hemolytic anemia and death may follow.

The treatment of choice in copper poisoning or Wilson's disease is oral administration of penicillamine. It can be administered orally in capsules. The immediate and most dramatic effect of the administration of penicillamine is a marked increase in urinary copper excretion. However, there are some undesirable side effects. With the use of triethylenetetramine dihydrochloride, marked cupriuria and clinical improvement in individuals with Wilson's disease have been achieved. The modes of action for these two chelating agents are apparently different.

Iron

Iron is an essential component of several cofactors, including hemoglobin and the cytochromes. There is an active complicated homeostatic mechanism for maintaining proper iron levels in the body. In this mechanism the divalent form is absorbed into the gastrointestinal mucosa. The adequacy of iron stores in the body, however, seems to be the major controlling factor in the absorption of iron by the gastrointestinal tract. Absorption increases when iron stores are low and decreases when body stores are sufficient.

Acute pulmonary exposures to iron compounds are associated with kidney and liver damage. Altered respiratory rates and convulsions are among the neurological effects. Roentgenological changes in the lungs, referred to as siderosis, iron or hematite pneumoconiosis, or arc welders lung occur following inhalation of iron oxide fumes. These changes are similar to those seen in silicosis or miliary tuberculosis.

Chronic exposure to or excessive intake of iron may lead to hemosiderosis or hemochromatosis. Hemosiderosis refers to a condition in which there is generalized increased iron content in the body tissues, particularly the liver and reticuloendothelial system. Hemochromatosis, on the other hand, indicates demonstrable histologic hemosiderosis and diffused fibrotic changes of the affected organ.

Desferrioxamine is the chelating agent of choice in the treatment of iron poisoning. It complexes with iron ions to form ferrioxamine which is then excreted via the kidney. One gram of desferrioxamine binds 85 mg of iron.

Lead

The essentiality of lead has been proposed at different times but it has never been fully demonstrated to the satisfaction of the scientific community. The symptoms of lead poisoning were noted by Greek, Roman, and Arabian physicians long before they were ascribed to lead. Lead "colic" was described by Hippocrates in a man who extracted metals and the relationship of constipation colic, pallor, and paralysis was noted by Nicander in the second century. By the early part of this century, the clinical signs and symptoms of lead toxicity were documented in considerable detail. Nonspecific signs and symptoms include loss of appetite, metallic taste in the mouth, constipation and obstipation, pallor, malaise, weakness, insomnia, headache, irritability, muscle and joint pains, fine tremors, and colic.

The central nervous system is probably the most clinically significant site of toxicity. Symptoms vary from ataxia to stupor, coma, and convulsions. Lead encephalopathy is the term commonly used to describe the damage to the brain by lead. Peripheral neuropathy, characterized by wrist drop and foot drop, is the classical manifestation of lead toxicity.

Lead has multiple hematopoietic effects. In lead-induced anemia, the red blood cells are microcytic and hypochromic as in iron deficiency and usually there are increased numbers of reticulocytes with basophilic stippling. Two basic defects of lead-induced anemia are shortened life span for red blood cells and impairment of heme synthesis. As a consequence of the latter, there is a marked increase in circulating blood levels and urinary excretion of δ-aminolevulinic acid. Lead also decreased ferrochelatase activity. This enzyme catalyzes the incorporation of the ferrous ion into the porphyrin ring structure. Failure to insert iron into protoporphyrin results in depressed heme formation. The excess protoporphyrin takes the place of heme in the hemoglobin molecule and zinc is chelated at the center of the molecule at the site usually occupied by iron. Red blood cells containing zinc protoporphyrin are intensely fluorescent and may be used to diagnose lead toxicity.

In workers with years of exposure to lead, toxicological effects on the kidney are most often observed. In the early stages, morphological and functional changes in the kidney are confined to the renal tubules and are most pronounced in proximal tubular cells. In advanced stages, irreversible chronic interstitial nephropathy characterized by vascular sclerosis, tubular cell atrophy, interstitial fibrosis, and glomerular sclerosis may occur. Gout occurs in about 50% of persons with chronic lead nephropathy.

Severe lead toxicity has long been known to cause sterility, abortion and neonatal mortality and morbidity. It has been found that female lead workers are abortifacient and that male lead workers have a high incidence of sterile marriages.

The possibility of carcinogenic effects of lead has been receiving increased attention. There is evidence that lead can induce cancer in kidneys of rodents when fed high doses of lead. However, there is currently no evidence that lead is carcinogenic to man.

The most commonly used chelating agent in the removal of lead from the body is the calcium salt of disodium ethylenediaminetetraacetate. The lead chelate formed by exchange of calcium for lead is promptly excreted in the urine. For treatment of children with lead poisoning, the statement issued by the center for Disease Control should be consulted.

Manganese

Manganese is an essential element. It is a cofactor in a number of enzymatic reactions, particularly those involved in phosphorylation, cholesterol, and fatty acids synthesis. Manganese toxicity has been observed primarily among workers associated with the mining, refining, and manufacturing of manganese. In these individuals, overt signs of toxicity normally occur as the result of chronic inhalation of massive amounts of airborne manganese. The initial expression of manganese toxicity is often characterized by a severe psychiatric disorder. If the individual is removed from the high-manganese environment, some improvement of the psychiatric symptoms of the toxicity may occur. With continued exposure, however, the symptoms may progress remarkably similar to those noted in Parkinson's disease. Individuals at this point tend to show mask-like faces, difficulty in walking, and exaggerated reflexes. Removal of the individual from the high-manganese environment at this point will not result in a remission of the disorder even though tissue manganese levels may decrease to normal values. An increase in respiratory diseases such as pneumonitis and pneumonia are frequently observed with milder forms of manganism.

In addition to the extensive neural tissue damage which can occur with chronic manganese toxicity, an iron-responsive anemia is a common finding with orally induced manganese toxicity. This anemia is presumably the result of an inhibitory effect of manganese on gastrointestinal uptake of iron. Reproduction and immune system dysfunction, nephritis, testicular damage, pancreatitis, and hepatic damage have all been reported.

Chelation theory of patients suffering from chronic manganese poisoning with penicillamine or ethylenediaminetetraacetate showed some improvement in patients. However, the improvements observed were transient. Currently, there is debate with regard to the efficiency of chelation therapy in the treatment of manganism as it has been reported that it is of little value in the treatment of this disorder. Support for this view is given by the observation that removal of an individual from a high-manganese environment results in a rapid loss of excess manganese from the body. Although body loss of manganese is reflected by lower concentrations of the element in the brain, lesions that occurred during exposure to the high-manganese are not reversible.

Mercury

Mercury is not an essential element in the body. Mercury and its compounds can be classified according to their dominant toxic characteristics into the following groups: (1) elemental mercury and those ionic compounds that can decompose to the vapor form in the environment; (2) inorganic mercury, e.g., both the mercury(I) and mercury(II) ions; (3) short-chain alkylmercurials such as methyl- and ethylmercury; and (4) organomercurials with more than two carbon atoms in the liquid, e.g., phenylmercury.

Exposure to a high concentration of mercury vapor can produce acute pneumonitis. Chronic low level exposure can increase the incidence of weight loss caused by loss of appetite. Higher levels can produce chronic mercurialism with the four classical signs of gingivitis, salivation, increased irritability, and muscular tremors. Proteinuria may also occur. The primary target organ of elemental mercury is the central nervous system. It is oxidized to the mercury(II) ion. This reaction limits, but does not prevent, the accumulation of mercury by the brain.

The first signs of the ingestion of inorganic mercury(II) salts are caused by a corrosive effect on the alimentary tract, followed by oliguria and acute renal failure. The prolonged ingestion of mercury(I) salts can cause an idiosyncratic disease characterized by irritability, loss of body weight, acrodynia (painful extremities), "pink" disease (rash), and photophobia. Neither the occurrence of the disease nor its severity is dose-related. The disposition of organomercurials with more than two carbon, atoms in the ligand is essentially the same as for inorganic mercury. This is because the carbon-mercury bond is rapidly cleaved *in vitro* yielding the mercury(II) ion and an organic moiety.

Short-chain alkylmercurials such as methylmercury primarily damages the central nervous system. The first symptom is paresthesia and the first clinical sign is ataxia followed in severe cases by the constriction of the visual field and deafness. Ethylmercury has the same effect on the central nervous system but signs of renal damage such as proteinuria are also present. All organomercurials can cause redness of skin, burning sensation, blisters, and hypersensitivity reactions after repeated exposure. Transplacental transport is significant for alkylmercury as well as for elemental mercury.

Chelating agents are used to remove mercury from the body. The first thiol chelating agent dimercaptopropanol, successfully decreased the mortality of acute mercury(II) intoxication, but the majority of the patients had adverse reactions to therapeutic doses. Dimercaptopropanol has two other drawbacks. It is ineffective when given orally and ineffective as a chelating agent against alkylmercury. The next major advance was the discovery that penicillamine was an effective chelating agent even when administered orally. More recently, N-acetyl-*dl*-penicillamine, 2,3-dimercaptopropanesulfonate, and 2,3-dimercaptosuccinic acid have been discovered to be even more effective and less toxic than penicillamine. They are all water-soluble and can be given orally.

Nickel

Nickel is an essential trace element in chicks, rats, and swine but not in human. Occupational health hazards from exposure to nickel compounds fall into three major categories: hypersensitivity, cancer, and respiratory disorders.

Hypersensitivity to nickel, or nickel itch, is a common cause of allergic dermatitis. The cutaneous lesions of nickel itch begin as a papular erythema (redness) of the hands or other areas of skin that contact nickel. The lesions gradually become eczematous (weeping) and undergo lichenification (scaling) in the chronic state.

Nickel refining workers have had increased mortality rates from cancers of the lungs and nasal cavities, attributed to exposure to nickel compounds with low aqueous solubilities, such as nickel sulfide and nickel oxide. Increased risks of other malignant tumors, including cancers of the larynx, kidney, prostate, or stomach have been noted in some refinery workers.

Owing to its volatility, lack of strong odor, and propensity for inadvertent formation, nickel carbonyl is generally considered the most hazardous compound of nickel. Symptoms of exposure to nickel carbonyl are of two distinct types. The immediate symptoms of toxicity consist of headache and vomiting. These are relieved by fresh air. Delayed and severe symptoms may then develop insidiously hours or even days after exposure. These symptoms include dyspnea, tachycardia, cyanosis, headache, dizziness, and profound weakness. Adrenal, hepatic, and renal damage may also develop. Diffuse interstitial pneumonitis and cerebral hemorrhage or edema are the usual causes of death.

The chelating agent of choice for treatment of acute nickel poisoning is penicillamine since it greatly promotes the renal excretion of nickel. In nickel carbonyl poisoning diethyldithiocarbamate and tetraethylenepentamine are the most effective chelating agents if administered immediately after inhalation. The therapeutic efficacy of these chelating agents reflect their ability to form an intracellular nickel complex with gradual release of the nickel complex into extracellular fluids. Penicillamine and triethylenetetramine are effective in the extracellular complexation of nickel and thus enhance its renal clearance. Thus, if treatment of nickel carbonyl poisoning is administered several hours after exposure, penicillamine or triethylenetetramine would be the chelating agents of choice.

Selenium

Selenium is an essential nutrient. Its best known biochemical function is as a component of the enzyme glutathione peroxidase. The enzyme prevents oxidative damage of important cell constituents.

The first and most characteristic sign of selenium absorption is a garlic odor of the breath of small amounts of dimethyl selenide. In selenium intoxication, the main symptoms are brittle hair with intact follicles, and pigmentless new hair as well as brittle nails with spots of longitudinal streaks on the surface. Damaged nails are replaced by new, thickened nails, rough on the surface. In more severe cases, fluids effuse from around the nail bed. Another common symptom is lesions on the skin, mainly on the backs of hands and feet, the outer side of the legs, and forearms, and the neck. Affected skin becomes red and swollen. Allergy to selenium may cause a pink discoloration of the eyelids and palpebral conjunctivitis ("rose-eye").

In addition to the apparent protective effect against some carcinogenic agents, selenium is an antidote to the toxic effects of certain metals. At appropriate concentrations, mutual detoxification of selenium and arsenic, selenium and cadmium, selenium and copper, selenium and mercury, and selenium and thallium has been demonstrated. However, the mechanisms underlying these interactions are unknown.

In treatment of selenium intoxication, a combination of dimercaptopropanol and vitamin C was found to be an effective antidote. Symptomatic treatment with oxygen has been recommended as a treatment for selenium oxide and hydrogen selenide inhalation. Painful skin, nail, and eye disorders caused by selenium have been successfully treated with solutions or ointments of thiosulfate.

Thallium

Thallium is not an essential element in the body. Whether acute or chronic, the most characteristic features in thallotoxicosis are those involving the nervous system, skin, and cardiovascular tract. The usual patterns of damage in the peripheral nervous system are those of the "dying-back" type, with some involvement of the central nervous system. Hair loss generally occurs during thallium poisoning. Cutaneous effects may also include dry, scaly skin and impairment of nail growth often resulting in the appearance of Mee's lines. Thallium poisoning may also produce a complex pattern of cardiovascular responses. Small doses of thallium have been reported to produce increases in heart rate and blood pressure but large doses may produce hypotension and bradycardia. Cardiovascular symptoms are often accompanied by retrosternal pain or electrocardiogram abnormalities such as flattening or inversion of the T-wave.

Current treatment of thallium poisoning is primarily directed to the elimination of the metal from the body. Treatment with Prussian blue, potassium ferric hexacyanoferrate(II), can accelerate the fecal elimination of thallium by forming a stable, unabsorbable compound in the intestinal tract. Other antidotes such as diethyldithiocarbamate and diphenyl thiocarbazone are contraindicated because they form lipophilic chelates with thallium which are redistributed to central nervous system structures and may cause more neurological damage.

Tin

Tin is not an essential element in the body. In general, metallic tin and inorganic tin compounds have low toxicity. Inhalation of tin dioxide over a number of years may lead to stannosis, a pneumoconiosis with benign characteristics. Tin hydride, however, is highly toxic. Its effect is comparable to that of arsine.

In contrast to the inorganic tin compounds, some of the organic compounds are strongly toxic. Those with short alkyl groups such as methyl and ethyl groups are particularly dangerous. They are derived from tetravalent tin. Their toxicity increases with the number of alkyl groups. Trimethyl tin chloride and triethyl tin chloride deserve particular attention since they are highly neurotoxic and cause neurological and psychiatric disorders.

Inorganic tin compounds are very poorly absorbed when taken orally. Organotin compounds are decomposed by hydrolysis, but this reaction proceeds only slowly. Currently, there is no known antidote for the treatment of tin poisoning.

Vanadium

Vanadium is an essential element for chicken, rats, and some plants but is currently not known to be essential in man. In occupational exposure, the upper respiratory tract is the main target. Vanadium compounds, especially vanadium pentoxide, are strong irritants of the eyes and the respiratory tract. Other respiratory effects of vanadium include allergic contact dermatitis, asthma, and a green discoloration of the tongue.

Eighteen chelating agents as antidotes for acute vanadium intoxication were investigated and most showed some activity. The most promising antidote in vanadium poisoning is ascorbic acid.

Zinc

Zinc is an essential element in the body. Zinc compounds are not very toxic to humans. Zinc oxide has posed perhaps the greatest toxic risk and then only from inhalation of zinc oxide fumes in an industrial setting. Reports of zinc toxicity due to occupational exposure are very rare. In fact, lack of adequate zinc intake appears to be a far greater threat to human health than overexposure.

The exceedingly low toxicity and tightly regulated absorption and metabolism of zinc have made active measures directed toward detoxification unnecessary. The nausea, vomiting, and diarrhea that accompany acute exposure are self-limiting, and symptoms associated with metal fume fever characteristically abate within hours.

REFERENCES

Fergusson, J. E., *The Heavy Elements: Chemistry, Environmental Impact and Health Effects*, Pergamon Press, New York, 1990.

Fishbein, L. and Furst, A. (eds), *Biological Effects of Metals*, Plenum Press, New York, 1987,

Klaassen, C. D. (ed), *Cassarett and Doull's Toxicology: The Basic Science of Poisons*, 5th edition, Pergamon Press, New York, 1995.

Seiler, H. G., Sigel, H., and Sigel, A., *Toxicity of Inorganic Compounds*, Marcel Dekker, Inc., New York, 1988.

Waldron, H. A. (ed), *Metals in the Environment*, Academic Press, New York, 1980.

3

Solvent Exposure and Toxic Responses

Stephen K. Hall

CONTENTS

INTRODUCTION

A solvent is any relatively nonreactive substance, usually a liquid at room temperature, that dissolves another substance, resulting in a solution. Since most of the substances that solvents are used to dissolve in industry are organic, most industrial solvents are organic chemicals. Solvents are employed in a

wide variety of applications, including as dry-cleaning agents, chemical intermediates, degreasers, and liquid extracts. There is a wide range in the ability
of solvents to dissolve a given substance, and a similar range in their toxicities
and relative hazards to potentially exposed workers. Exposure to solvents
occurs primarily through inhalation of vapors and through skin contact.

Solvents may cause toxic effects in an exposed individual. They affect
several organ systems. The central and peripheral nervous systems are particularly susceptible, with effects ranging from slight decreases in nerve conduction velocity to narcosis and death. The acute neurological effects are related
to the anesthetic property of organic solvents, manifesting as transient symptoms such as lightheadedness and dizziness. Chronic neurological effects may
include loss of intellectual function and memory. The blood, lungs, liver,
kidneys, and skin also may be adversely affected by exposure to a particular
solvent. Classically, the halogenated hydrocarbons are capable of inducing
fatty changes and cirrhosis of the liver. Dermatitis, a common result of prolonged or repeated contact with solvents, is due primarily to defatting of the
skin tissues. Selected solvents have been related to the destruction of the bone
marrow. Other solvents are known to cause cardiovascular effects, renal effects,
as well as other health effects in the exposed individual. Some solvents are
known human carcinogens; others are animal carcinogens suspected of possessing carcinogenic activity in humans.

METABOLISM OF ORGANIC SOLVENTS

Organic solvents can be divided into families according to chemical structure
and the attached functional groups. Toxicological properties tend to be similar
within a group, such as liver toxicity from chlorinated hydrocarbons and
respiratory tract irritation from ketones. The basic structures are aliphatic,
alicyclic, and aromatic hydrocarbons. The functional groups include halogens,
alcohols, ketones, ethers, esters, amines, and others.

Since organic solvents are generally volatile liquids, inhalation is the primary route for occupational exposure. The pulmonary retention or uptake for
most organic solvents ranges from 40 to 80% at rest. Because physical labor
increases pulmonary ventilation and blood flow, the amount of organic solvent
delivered to the alveoli and the amount absorbed are likewise increased.

Upon direct contact with the skin, the lipid solubility of organic solvents
results in most being absorbed through the skin. However, skin absorption is
also determined by water solubility and volatility. Solvents that are soluble in
both water and lipid are most readily absorbed through the skin. Highly volatile
substances are less well absorbed. For a number of solvents, dermal absorption
contributes to overall exposure sufficient to result in a "skin" notation for the
Threshold Limit Values (TLVs) of the American Conference of Governmental
Industrial Hygienists (ACGIH). For a few solvents, significant absorption of
vapors through the skin can also occur.

After absorption, organic solvents tend to be distributed to fatty tissues. In addition to adipose tissue, this includes the nervous system and liver. Since distribution occurs via the blood and since the blood-tissue membrane barriers are usually rich in lipids, solvents are also distributed to organs with large blood flows, such as cardiac and skeletal muscle. Persons with greater amounts of adipose tissue will accumulate greater amounts of an organic solvent over time. Most organic solvents will cross the placenta and also enter breast milk.

Some organic solvents are extensively metabolized and some not at all. The metabolism of a number of organic solvents plays a key role in their toxicity and in some cases the treatment of intoxication. The role of toxic metabolites will be discussed in their respective sections in this chapter. Biological monitoring can provide a more accurate measure of exposure than environmental monitoring for some solvents; this important topic will be discussed and summarized in a later chapter.

Excretion of organic solvents occurs primarily through exhalation of unchanged compound, elimination of metabolites in urine, or a combination of both. Solvents that are poorly metabolized are excreted primarily through exhalation.

Following are summaries of toxic effects of organic solvents according to their chemical family and functional group. These solvents were selected for discussion because of their economic value, industrial use, and documentation of their toxic effects on exposed individuals. Each summary contains the clinical manifestation of signs and symptoms of toxicity, target organs or systems, metabolism, and carcinogenicity. Wherever appropriate, specific solvents within a chemical family are discussed.

Aliphatic Hydrocarbons

Aliphatic hydrocarbons are saturated or unsaturated, branched or unbranched, open carbon chains. They are further divided into alkanes (saturated hydrocarbons with carbon-to-carbon single bonds), alkenes (unsaturated hydrocarbons with one or more carbon-to-carbon double bonds), and alkynes (unsaturated hydrocarbons with one or more carbon-to-carbon triple bonds). Synonyms are paraffins, olefins, and acetylenes, respectively. Compounds of lower molecular weight containing fewer than four carbons are gases at room temperature, whereas larger molecules containing from five to sixteen carbons, are liquids, and those having more than sixteen carbons are usually solids.

A number of liquid aliphatic hydrocarbons are used in relatively pure form as solvents and also are the major constituents of a number of petroleum distillate solvents. The liquid alkanes are important ingredients in gasoline, which accounts for most of the pentane and hexane used worldwide. Hexane is an inexpensive general-use solvent in solvent glues, quick-drying rubber cements, varnishes, inks, and extraction of oils from seeds.

The alkanes are generally of low toxicity. The first three alkanes (methane, ethane, and propane) are simple inert asphyxiants whose toxicity is related only to the amount of available oxygen remaining in the environment. The vapors of the lighter, more volatile liquids, pentane through nonane, are respiratory tract irritants and anesthetics, while the heavier liquids, known as liquid paraffins, are primarily defatting agents. Hexane and heptane are most commonly used as general purpose solvents. They cause anesthesia, respiratory tract irritation, and dermatitis and are associated with neurobehavioral dysfunction.

n-Hexane is produced during the cracking and fractional distillation of crude oil and is used in such applications as printing of laminated products; vegetable oil extraction; as a solvent in glues, paints, varnishes, and inks; as a diluent in the production of plastics and rubber; and as a minor component of gasoline. Vapor concentrations of many hundreds of parts per million are tolerated for several minutes without causing discomfort among workers. Peripheral neuropathy was reported in workers involved in laminating polyethylene products. The proximate neurotoxin is the metabolite 2,5-hexanedione, which is excreted in the urine. Methyl ethyl ketone (MEK) and possibly methyl isobutyl ketone (MiBK) potentiate the neurotoxicity of *n*-hexane.

The liquid alkenes are not widely used as solvents but are common chemical intermediates. They are more reactive than alkanes, a property that leads to their use as monomers in the production of polymers, such as polyethylenes from ethylene, polypropylene from propylene, and synthetic rubber and resin copolymers from 1,3-butadiene. The alkenes are similar in toxicity to the alkanes. The double bonds increase lipid solubility and therefore irritant and anesthetic potencies, compared to corresponding alkanes. *n*-Hexene does not cause peripheral neuropathy, as does *n*-hexane. Dienes are more reactive than alkenes. This reactivity is utilized in the production of polymers but may in some cases result also in additional health hazards. 1,3-Butadiene was found to be carcinogenic in animals, while propylene and ethylene were not.

Alicyclic Hydrocarbons

Alicyclic hydrocarbons are saturated or unsaturated molecules in which three or more carbon atoms are joined to form a ring structure. The saturated compounds are called cycloalkanes, cycloparaffins, or naphthenes. The cyclic hydrocarbons with one or more double bonds are called cycloalkenes or cycloolefins. Cyclohexane is the only alicyclic hydrocarbon that is widely used as an industrial solvent for fats, oils, waxes, resins, and certain synthetic rubbers, and as an extractant of essential oils in the perfume industry. However, most of the cyclohexane produced is used in the manufacturing of nylon. Cyclohexene is used in the manufacture of adipic, maleic, and cyclohexane carboxylic acid. Methylcyclohexane is used as a solvent for cellulose ethers and in the production of organic synthetics.

The aliphatic hydrocarbons are similar in toxicity to their alkane or alkene counterparts in causing respiratory tract irritation and central nervous system depression, although their acute toxicity is low. The danger of chronic poisoning is relatively slight because these compounds are almost completely eliminated from the body. Alicyclic hydrocarbons are excreted in the urine as sulfates or glucuronides, the particular content of each varying. Small quantities of these compounds are not metabolized and may be found in blood, urine, and expired breath.

Aromatic Hydrocarbons

Aromatic hydrocarbons are characterized by the presence of the aromatic nucleus. The basic aromatic nucleus is the benzene ring. Aromatic hydrocarbons have enjoyed wide usage as solvents and as chemical intermediates. They are produced chiefly from crude petroleum and to a lesser extent from coal tar. Aromatic hydrocarbons used as solvents include benzene, toluene (methyl benzene), xylene (dimethyl benzene), ethyl benzene, cumene (isopropyl benzene), and styrene (vinyl benzene). They have a characteristic aromatic (sweet) odor.

Benzene is still currently widely used in manufacturing, for extraction in chemical analyses, and as a specialty solvent. Approximately half the benzene produced is used to synthesize ethyl benzene for the production of styrene. Toluene and the xylenes are two of the most widely used industrial solvents used in paints, adhesives, and the formulation of pesticides. Ethyl benzene is used chiefly as an intermediate in the manufacture of plastics and rubber. Cumene is used to manufacture phenol and acetone.

All the aromatic hydrocarbons are extensively metabolized. The metabolites vary with the substituents on the benzene ring. Benzene is metabolized mainly to phenol and excreted in urine as conjugated phenol and dihydroxyphenols. Toluene is primarily metabolized to benzoic acid and excreted in urine as the glycine conjugate hippuric acid. Xylenes are metabolized to methylbenzoic acids and excreted in urine as the glycine conjugates methylhippuric acid. Ethyl benzene is metabolized and excreted in urine as mandelic acid. Styrene is metabolized and excreted in urine primarily as mandelic acid and, to a lesser extent, phenylglyoxylic acid.

The aromatic hydrocarbons are generally stronger respiratory tract irritants and anesthetics than the aliphatics containing the same number of carbon atoms in the molecule. Substitution on benzene (toluene, xylene, ethyl benzene, cumene, and styrene) increases lipid solubility and toxicities slightly. As a family, aromatic hydrocarbons cause acute anesthetic effects, dermatitis and respiratory tract irritation, and are associated with neurobehavioral dysfunction. Benzene is noted for its effects on the bone marrow: aplastic anemia that may itself be fatal or progress to leukemia. There is no evidence that any of the alkyl-substituted benzenes have any of the myelotoxic effects. Earlier reports of effects of the substituted benzenes on the bone marrow were probably due to their contamination with benzene.

Petroleum Distillates

Petroleum distillate solvents are mixtures of petroleum derivatives distilled from crude petroleum at a particular range of boiling points. Each is a mixture of aliphatic, alicyclic, and aromatic hydrocarbons, the relative concentration of each depending on the particular petroleum distillate fraction. They have a "hydrocarbon" or "aromatic" odor depending on the relative concentrations of aliphatic or aromatic hydrocarbons. "Petroleum ether" represents more than half of the total industrial solvent usage. Kerosene (stove oil) is used as a fuel as well as a cleaning and thinning agent. In industry, petroleum distillates may be referred to by any of the following names: naphtha, coal tar naphtha, petroleum naphtha, mineral spirits, Stoddard solvent, and others.

The hazard of a particular petroleum distillate fraction is related to concentrations of the various classes of hydrocarbons it contains. Most of the aliphatic fractions are alkanes, including n-hexane. Therefore, the risk of peripheral neuropathy must be considered, particularly with exposure to petroleum ether, which may contain a significant percentage of n-hexane. As the fraction becomes heavier (higher boiling point, increasing number of carbons), the percentage of aromatic hydrocarbons and therefore the toxicity increases. However, the increase in toxicity is offset by a decrease in volatility. Petroleum distillate solvents cause anesthetic effects, respiratory tract irritation and dermatitis, and have been associated with neurobehavioral dysfunction.

Chlorinated Hydrocarbons

The addition of chlorine to a hydrocarbon increases the stability and decreases the flammability of the resulting compounds. Chlorinated hydrocarbons are typically colorless, volatile liquids with excellent solvent properties. Chemically, they consist of saturated or unsaturated carbon chains in which one hydrogen atom or more has been replaced by one chlorine atom or more. Hydrocarbons having only one or two chlorine are usually flammable and less toxic than similar hydrocarbons with complete chlorine substitution. Chlorinated hydrocarbon solvents are moderately well absorbed by inhalation. Skin absorption of vapor is usually insignificant, but skin absorption following prolonged contact of the skin with liquid can be significant.

Chlorinated hydrocarbons have found wide use as solvents in degreasing, dewaxing, dry-cleaning, and extracting processes. Six chlorinated aliphatic hydrocarbons are used as general industrial solvents: methylene chloride, chloroform, carbon tetrachloride, methyl chloroform, trichloroethylene, and perchloroethylene. Methylene chloride is used as a paint stripper and extraction agent. Chloroform is used for extraction and spot-cleaning. Carbon tetrachloride is used primarily as a chemical intermediate and in small quantities as a spot cleaning agent. Methyl chloroform is used in vapor degreasers and increasingly as a general cleaning and thinning agent. Historically, trichloroethylene was the principal solvent used in vapor degreasers; it is being replaced by methyl chloroform, which is less toxic. Perchloroethylene has replaced mineral spirits and carbon tetrachloride as the primary dry-cleaning solvent because of the flammability of the former and the toxicity of the latter.

Biological monitoring of the chlorinated hydrocarbon is based on their pattern of metabolism and excretion, which varies with their structure. Methylene chloride is both excreted unchanged in exhaled air and metabolized to carbon monoxide. Chloroform and carbon tetrachloride are each excreted unchanged in exhaled air and metabolized. However, little information is available on biologic monitoring for either. Methyl chloroform and trichloroethylene, and perchloroethylene are excreted unchanged in exhaled air and metabolized and excreted as trichloroethanol and trichloroacetic acid.

Alcohols

Alcohols are hydrocarbons substituted with a single hydroxyl group. They are widely used as cleaning agents, thinners, and diluents; as vehicles for paints, pesticides, and pharmaceuticals; as extracting agents; and as chemical intermediates. Methyl alcohol is widely used as an industrial solvent and as an adulterant to denature ethanol to prevent its abuse when used as an industrial solvent. Another important industrial use of methyl alcohol is in the production of formaldehyde. Isopropyl alcohol is used as rubbing alcohol and in the manufacture of acetone. Cyclohexanol is used to produce adipic acid for the production of nylon. In general, the aliphatic alcohols with more than five carbon atoms are divided into the plasticizer range (6 to 11 carbons) and the detergent range (>12 carbons).

Low-molecular weight alcohols are volatile and up to 50% of the inhaled vapor of these low-molecular weight alcohols can be absorbed into the body. In addition, vapors of methyl alcohol, isopropyl alcohol, *n*-butyl alcohol, and *iso*-octyl alcohol can be absorbed through the skin and they carry the "skin" TLV notation.

The primary alcohols are metabolized by hepatic alcohol dehydrogenase to aldehydes and by aldehyde dehydrogenase to carboxylic acids. The optic neuropathy and metabolic acidosis caused by methyl alcohol have been attributed to its metabolism to formaldehyde and then to formic acid. Metabolic interactions of ethanol with chlorinated hydrocarbon solvents, such as "degreaser's flush" in workers exposed to trichloroethylene are due to competition for alcohol and aldehyde dehydrogenases, with subsequent accumulation of the alcohol and aldehyde and resulting reaction. Secondary alcohols are primarily metabolized to ketones (e.g., isopropyl alcohol is metabolized to acetone).

The aliphatic alcohols are more potent central nervous system depressants and irritants than the corresponding aliphatic hydrocarbons, but they are weaker skin and respiratory tract irritants than aldehydes or ketones. Respiratory tract and eye irritation usually occur at lower concentrations than central nervous system depression and thus serve as a useful warning property.

Methyl alcohol is toxicologically distinct and targets the optic nerve, which can result in blindness. Inhalation exposure to ethanol and propanols result in simple irritation and central nervous system depression. Auditory and vestibular nerve injury in workers exposed to *n*-butyl alcohol has been reported.

Glycols

Glycols are hydrocarbons with two hydroxyl groups attached to two adjacent carbon atoms in an aliphatic chain. Glycols are used as antifreezing agents and as solvent carriers and vehicles in a variety of chemical formulations. Only ethylene glycol is in common general industrial use as a solvent, but large volumes of the other glycols are used as vehicles and chemical intermediates.

The glycols have such low vapor pressures that inhalation is only of moderate concern unless heated or aerosolized. Ethylene glycol and diethylene glycol are metabolized to glycol aldehyde, glycolic acid, glyoxylic acid, oxalic acid, formic acid, glycine, and carbon dioxide. Oxalic acid is the cause of acute renal failure and metabolic acidosis that occur following ingestion of ethylene glycol. Urinalysis for oxalic acid may be useful in biological monitoring of ethylene glycol exposure. The metabolic pathways of methyl alcohol and ethylene glycol may be competitively blocked by the administration of ethyl alcohol.

Phenols

Phenols are aromatic hydrocarbons with one or more hydroxyl groups attached to the benzene ring. The simplest of the compounds is phenol, which contains only one hydroxyl group on a benzene ring. Other examples include cresol (methyl phenol), catechol (1,2-benzenediol), resorcinol (1,3-benzenediol), and hydroquinone (1,4-benzenediol). Phenol is used as a cleaning agent and disinfectant, but its primary use is as a chemical intermediate for resins and pharmaceuticals. Cresol is used primarily as a disinfectant. Catechol is used in photography, fur dying, leather tanning, and as a chemical intermediate. Resorcinol is used as a chemical intermediate for adhesives, dyes, and pharmaceuticals. Hydroquinone is used in photography, as a polymerization inhibitor, and as an antioxidant.

Phenol is well absorbed both by inhalation of vapors and by dermal penetration of vapors and liquids. Phenol and cresol have "skin" TLV notations. Phenols are potent irritants that can be corrosive at high concentrations. As a result of their ability to complex with, denature, and precipitate proteins, they can be cytotoxic to all cells at sufficient concentrations. Direct contact with concentrated phenol can result in burns, local tissue necrosis, systemic absorption, and tissue necrosis in the liver, kidneys, urinary tract and heart. As it does with aliphatic alcohols and other volatile organic solvents, central nervous system depression also occurs.

Ketones

Ketones are hydrocarbons with a carbonyl group attached to a secondary carbon atom. Ketones are widely used as solvents for surface coatings with natural and synthetic resins; in the formulation of inks, adhesives, and dyes; in chemical extraction and manufacture; and as cleaning agents. Acetone, methyl ethyl ketone (MEK), and cyclohexanone are in most common use as industrial solvents. Consumer exposure to acetone is common in the form of nail polish remover and general use solvent. Acetone is also used in the manufacture of methacrylates while cyclohexanone is used to make caprolactam for nylon.

Ketones are well absorbed by inhalation of vapors and to some extent after skin contact with liquid. Cyclohexanone has a "skin" TLV notation. Most ketones are rapidly eliminated unchanged in urine and exhaled air and by reduction to their respective secondary alcohols, which are conjugated and excreted or further metabolized to a variety of compounds, including carbon monoxide.

Ketones are colorless liquids with good warning properties; a strong odor or irritation to the skin, eyes, and respiratory tract usually occurs at levels below those that cause central nervous system depression. Headaches and nausea as a result of the odor have been mistaken for central nervous system depression.

Methyl *n*-butyl ketone (MnBK), or 2-hexanone, is another hexacarbon solvent such as *n*-hexane that was thought to have little potential for health hazard. It is used as a paint thinner, cleaning agent, solvent for dye printing, and in the lacquer industry. Methyl *n*-butyl ketone causes the same type of peripheral neuropathy as *n*-hexane. It is also metabolized to the neurotoxic diketone, 2,5-hexanedione, to an even greater extent than *n*-hexane and therefore poses an even greater hazard.

Ethers

Ethers consist of two hydrocarbon groups joined by an oxygen linkage. The two commonly used ethers as industrial solvents are ethyl ether and dioxane. Ethyl ether, the simplest ether, has been used extensively in the past as a general anesthetic, and has been historically known as "ether". It has now been replaced by agents less flammable and with less aftereffects. Ethyl ether is used as a solvent for waxes, fats, oils, and gums. Dioxane is used as a solvent for a wide range of organic products, including cellulose esters, rubber, and coatings; in the preparation of histologic slides, and as a stabilizer in chlorinated solvents.

Ethyl ether is well absorbed by inhalation of vapors. Its volatility limits skin absorption. Absorbed ethyl ether is excreted unchanged in exhaled air. The rest may be metabolized by enzymatic cleavage of the oxygen link to acetaldehyde and acetic acid. Ethyl ether is a potent anesthetic. Higher ethers are relatively more potent irritants. Dioxane is both an anesthetic and irritant but has also caused acute kidney and liver necrosis in workers. It is well absorbed by inhalation of vapors and through skin contact with liquid and has a "skin" TLV notation. Dioxane is metabolized almost entirely to β-hydroxyethoxyacetic acid and excreted in urine.

Occupationally, exposure to chlorinated ethers, bis(chloromethyl) ether (BCME) and chloromethyl methyl ether (CMME), is much more significant. Bis(chloromethyl) ether is used as an alkylating agent in the manufacture of polymers, as a solvent for polymerization reactions, in the preparation of ion exchange resins, and as an intermediate for organic synthesis. Bis(chloromethyl) ether has an extremely suffocating odor even in minimal concentrations. Bis(chloromethyl) ether is a known human carcinogen. There have been several reports of increased incidence of human lung carcinomas among workers exposed to bis(chloromethyl) ether as an impurity. The latency period is relatively short—10 to 15 years. Smokers as well as nonsmokers may be affected.

Chloromethyl methyl ether is a highly reactive methylating agent and is used in the chemical industry for synthesis or organic chemicals. Commercial grade chloromethyl methyl ether contains from 1 to 7% bis(chloromethyl) ether. Several studies of workers with chloromethyl methyl ether manufacturing exposure have shown an excess of bronchogenic carcinoma. It is not known whether or not the carcinogenic activity of chloromethyl methyl ether is due to bis(chloromethyl) ether contamination, but this may be a moot question inasmuch as two of the hydrolysis products of chloromethyl methyl ether can combine to form bis(chloromethyl) ether.

Esters

Esters are hydrocarbons that are derivatives of an organic acid and an alcohol. They are named after their parent alcohols and organic acids respectively, e.g., ethyl acetate for the ester of ethyl alcohol and acetic acid. The organic acid may be aliphatic or aromatic and may contain other substituents. Esters are an industrially important group of compounds. They are used in plastics and resins, as plasticizers, in lacquer solvents, in flavors and perfumes, in pharmaceuticals, and in industries such as automotive, aircraft, food processing, chemical pharmaceutical, soap, cosmetic, surface coating, textile, and leather.

Many esters have extremely low odor thresholds. Their distinctive sweet smells serve as good warning properties. Because of this property, n-amyl acetate (banana oil) is used as an odorant for qualitative fit testing of respirators. Esters are very rapidly metabolized by plasma esterases to their parent organic alcohols and acids. In general, esters are more potent anesthetics than corresponding alcohols, aldehydes, or ketones but are also strong respiratory tract irritants. Odor and irritation usually occur at levels below central nervous system depression. Their systemic toxicity is determined to a large extent by the toxicity of the corresponding alcohol.

There are four basic types of physiological effects of esters, and these can generally be related to structure. (1) Anesthesia and primary irritation are characteristic of most simple aliphatic esters. (2) Lacrimation, vesication, and lung irritation are due to the halogen atom in halogenated esters. (3) Cumulative organic damage to the nervous system or neuropathy can be caused by some, but not all, phosphate esters. (4) Most aliphatic and aromatic esters used as plasticizers are physiologically inert.

Glycol Ethers

Glycol ethers are alkyl ether derivatives of ethylene, diethylene, triethylene, and propylene glycol. The acetate derivatives of glycol ethers are included and are considered toxicologically identical to their precursors. They are known by formal chemical names, e.g., ethylene glycol monomethyl ether; common chemical names, e.g., 2-methoxyethanol; and trade names, e.g., methyl cellosolve.

The glycol ethers are widely used solvents because of their solubility or miscibility in water and most organic liquids. They are used as diluents in paints, lacquers, enamels, inks, and dyes; as cleaning agents in liquid soaps, dry-cleaning fluids, and glass cleaners; as surfactants, fixatives, desiccants, antifreeze agents, and deicers; and in extraction and chemical synthesis. Since the first two members of this family, methyl cellosolve and ethyl cellosolve, were found to be potent reproductive toxins in laboratory animals, there has been a shift in use to butyl cellosolve and to diethylene and propylene glycol ethers.

The glycol ethers are well absorbed by all routes of exposure owing to their universal solubility. The acetate derivatives are rapidly hydrolyzed by plasma esterases to their corresponding monoalkyl ethers. The ethylene glycol monoalkyl ethers maintain their ether linkages and are metabolized by hepatic alcohol and aldehyde dehydrogenases to their respective aldehyde and acid metabolites. The acid metabolites of 2-methoxyacetic acid and 2-ethoxyacetic acid are responsible for the reproductive toxicities of 2-methoxyethanol and 2-ethoxyethanol. These metabolites are excreted in urine unchanged or conjugated to glycine and may be used as biologic exposure indices.

Cases of encephalopathy have been reported in workers exposed to 2-methoxyethanol over long periods of weeks to months. Manifestations have included personality changes, memory loss, difficulty in concentrating, lethargy, fatigue, loss of appetite, weight loss, tremor, gait disturbances, and slurred speech. Bone marrow toxicity manifested as pancytopenia has been reported in workers exposed to 2-methoxyethanol and 2-ethoxyethanol.

Male reproductive toxicity has been observed in laboratory animals for 2-methoxyethanol and 2-ethoxyethanol in reduction of sperm count, impaired sperm motility, increased numbers of abnormal forms, and infertility. The same glycol ethers that are testicular toxins have been shown to be teratogenic in the same species of laboratory animals.

Glycidyl Ethers

The glycidyl ethers consist of a 2,3-epoxypropyl group with an oxygen linkage to another hydrocarbon group. They are synthesized from epichlorohydrin and an alcohol. Only the monoglycidyl ethers are in common use. The epoxide ring of glycidyl ethers makes these compounds very reactive, so their use is confined to processes that utilize this property such as reactive diluents in epoxy resin systems.

The glycidyl ethers have low vapor pressures, so that inhalation at normal air temperatures is not usually a concern. However, the curing of epoxy resins often generates heat, which may vaporize some glycidyl ether. Reported effects of glycidyl ether exposure have been confined to dermatitis of both the primary irritant and allergic contact type. Dermatitis can be severe and may result in second-degree burns. Asthma in workers exposed to epoxy resins may be due to exposure to glycidyl ethers. Glycidyl ethers are testicular toxins in laboratory animals.

Aliphatic Amines

The aliphatic amines are derivatives of ammonia in which one or more hydrogen atoms are replaced by a hydrocarbon group. They can be classified as primary, secondary, or tertiary according to the number of substitutions on the nitrogen atom. They are used to some extent as solvents but to a greater degree as chemical intermediates. They are also used as catalysts for polymerization reactions, preservatives (bactericides), corrosion inhibitors, drugs, and herbicides.

Amines are basic compounds and may form strongly alkaline solutions which can be highly irritating and cause damage on contact with eyes and skin. Amines are well absorbed by inhalation, and some have "skin" TLV notations. Skin absorption may be significant as many are capable of cutaneous sensitization. The vapors of the volatile amines cause eye irritation and a characteristic corneal edema, with visual changes of halos around lights, that is reversible. Irritation occurs wherever contact with the vapors occurs, including the respiratory tract and skin. Direct contact with the liquid can produce serious eye or skin burns. Allergic contact dermatitis has been reported from exposures to ethyleneamines.

Aromatic Amines

The aromatic amines are aromatic hydrocarbons in which at least one hydrogen atom has been replaced by an amino group. The hydrogen atoms in the amino group may be replaced by aryl or alkyl groups, giving rise to secondary and tertiary amino compounds. Their most important uses are as intermediates in the manufacture of dyestuffs and pigments; however, they are also used in the chemical, textile, rubber, dyeing, paper, and other industries.

Most of the aromatic amines in the free base form are readily absorbed through the skin in addition to the respiratory route. The two major toxic effects are methemoglobinemia and cancer of the urinary tract. Other effects may be hematuria, cystitis, anemia, and skin sensitization. Several of the aromatic amines have been shown to be carcinogenic in humans or animals or both. The most common site of cancer is the bladder, but cancer of the pelvis, ureter, kidney, and urethra do occur.

Aniline is a clear, colorless, oily liquid with a characteristic odor. It is widely used as an intermediate in the synthesis of dyestuffs. It is also used in the manufacture of rubber accelerators and antioxidants, pharmaceuticals, marking inks, tetryl, optical whitening agents, photographic developers, resins, varnishes, perfumes, shoe polishes, and many organic chemicals. Absorption of aniline, whether from inhalation of the vapor or from skin absorption of the liquid, causes anoxia due to the formation of methemoglobin. Moderate exposure may cause only cyanosis. As oxygen deficiency increases, the cyanosis may be associated with headache, weakness, irritability, drowsiness, dyspnea, and unconsciousness. Methemoglobin levels, and/or urinary excretion of p-aminophenols, can be used for biologic monitoring for aniline exposure.

Miscellaneous Solvents

Carbon disulfide.   Carbon disulfide is a highly refractive, flammable liquid which in pure form has a sweet odor and in commercial and reagent grades has a foul smell. It can be detected by odor at about 1 ppm but the sense of smell fatigues rapidly and therefore, odor does not serve as a good warning property. Carbon disulfide is used as a solvent for phosphorus, sulfur, selenium, bromine, iodine, alkali cellulose, fats, waxes, lacquers, camphor, resins, and cold vulcanized rubber. It is also used in degreasing, chemical analysis, electroplating, grain fumigation, oil extraction, and dry-cleaning.

Carbon disulfide vapor in sufficient quantities is severely irritating to eyes, skin, and mucous membranes. Inhalation of vapor may be compounded by percutaneous absorption of liquid or vapor. Contact with liquid may cause blistering with second- and third-degree burns. Skin sensitization may occur. Skin absorption may result in localized degeneration of peripheral nerves which is most often noted in the hands. Respiratory irritation may result in bronchitis and emphysema, though these effects may be overshadowed by systemic effects. Intoxication from carbon disulfide is primarily manifested by psychological, neurological, and cardiovascular disorders. Acute exposures may result in extreme irritability, uncontrollable anger, suicidal tendencies, and a toxic manic depressive psychosis. Chronic exposures have resulted in insomnia, nightmares, defective memory, and impotency. Less dramatic changes include headache, dizziness, and diminished mental and motor ability, with staggered gait and loss of coordination.

Atherosclerosis and coronary heart disease have been significantly linked to exposure to carbon disulfide. Atherosclerosis develops notably in the blood vessels of the brain, glomeruli, and myocardium. A significant increase in coronary heart disease mortality has been observed in carbon disulfide workers. Abnormal electrocardiograms may occur.

Carbon disulfide can be determined in expired air and blood. It is metabolized and excreted in the urine as 2-thiothiazolidine-4-carboxylic acid.

Dimethylformamide.   Dimethylformamide is a colorless liquid which is soluble in both water and organic solvents. It has a fishy, unpleasant odor at relatively low concentrations, but the odor has no warning property. Dimethylformamide has powerful solvent properties for a wide range of organic compounds. It finds particular usage in the manufacture of polyacrylic fibers, butadiene, purified acetylene, pharmaceuticals, dyes, petroleum products, and other organic chemicals. However, these properties also result in its being well absorbed by all routes of exposure.

Dimethylformamide is a potent hepatotoxin and has been associated with both hepatitis and pancreatitis following occupational exposure. It has produced kidney damage in animals. Recent studies have associated dimethylformamide exposure with testicular cancer. Exposure can be monitored biologically by measuring monomethylformamide and related metabolites in urine.

Turpentine. Turpentine is a mixture of substances called terpenes, primarily pinene. Gum turpentine is extracted from pine pitch, wood turpentine from wood chips. Inhalation of vapor and percutaneous absorption of liquid are the usual paths of occupational exposure. High vapor concentrations are irritating to the eyes, nose, and bronchi. Aspiration of liquid may cause direct lung irritation resulting in pulmonary edema and hemorrhage. Turpentine liquid may produce allergic contact dermatitis. The incidence of sensitization varies with the type of pine, being generally higher with European than American pines. Limonene is a terpene used as a solvent for art paints that also causes allergic contact dermatitis. Eczema from turpentine is quite common and has been attributed to the auto-oxidation products of the terpenes (formic acid, formaldehyde, and phenols). Liquid turpentine splashed in the eyes may cause corneal burns and demands emergency treatment.

Turpentine vapor in acute concentrations may cause central nervous system depression. It also produces kidney and bladder damage. Chronic nephritis with albuminuria and hematuria has been reported as a result of repeated exposures to high concentrations.

REFERENCES

Clayton, G. D. and Clayton, F. E. (eds), *Patty's Industrial Hygiene and Toxicology*, Volume 2, 4th edition, Wiley-Interscience, New York, 1994.

Commission of the European Communities, *Organo-Chlorine Solvents*, Royal Society of Chemistry, London, 1986.

Klaassen, C. D. (ed), *Cassarett and Doull's Toxicology: The Basic Science of Poisons*, 5th edition, Pergamon Press, New York, 1995.

Snyder, R., *Ethel Browning's Toxicity and Mechanism of Industrial Solvents*, Elsevier, New York, 1992.

Sullivan, Jr., J. B. and Krieger, G. R., *Hazardous Materials Toxicology*, Williams & Wilkins, Baltimore, 1992.

4

Chemical Exposure and Cancer

Gary D. Stoner

CONTENTS

INTRODUCTION

Cancer is the second leading cause of death in the United States, accounting for about 450,000 deaths each year, or approximately 20% of all deaths. Most cases are now believed to result from exposure to environmental chemicals.

These chemicals may be found in tobacco smoke, the diet, ambient air, drinking water, and in the workplace. The proportion of cancer deaths thought to be attributable to occupational exposure to carcinogens in the workplace has been estimated to be between 2 and 8%. Even if only 2% of all cancer deaths were attributed to occupational causes, this would amount to more than 8,000 deaths in the United States each year.

The first medical report of an occupational cancer was that of Percival Pott in 1775 describing skin cancer in chimney sweeps. Not until 1933, though, was the probable agent identified as benzo(a)pyrene, one of the many carcinogens found in tars. Since then, many substances and industrial processes have been proposed as causes of malignant disease, but fewer than 20 chemicals have been shown to be clearly carcinogenic in humans by means of epidemiologic studies in occupational settings. In this chapter, a review is presented of current knowledge regarding those chemicals known or suspected to be human carcinogens and encountered in the occupational setting. In addition, the chapter provides a discussion of the mechanisms of chemically induced cancer, and of the epidemiology and prevention of occupational cancer.

CHEMICAL CARCINOGENS IN OCCUPATIONAL SETTINGS

Table 4.1 lists chemical substances found in the workplace for which there is sufficient evidence of human carcinogenicity. The target organs in which these chemicals have been shown to induce cancer as well as the occupations of exposed workers is given. Additional comments regarding each of these chemicals or mixtures of chemicals follow.

4-Aminobiphenyl

This aromatic amine was formerly used as a rubber antioxidant and dye intermediate. The evidence for carcinogenicity of 4-aminobiphenyl to humans is confined to one series of workers exposed occupationally to commercial 4-aminobiphenyl; they showed a high incidence of bladder cancer.

Arsenic and Arsenic Trioxide Compounds

More than 90% of arsenic produced in the United States is a by-product of copper and lead ore smelting. Excess lung cancer has been reported in association with the use and production of inorganic trivalent arsenic-containing pesticides, as well as smelting operations. An estimated 545,000 workers are potentially exposed.

Asbestos Fibers

The four commercially important forms of asbestos are chrysotile, amosite, anthophylite, and crocidolite. Chrysotile represented 94% of United States

TABLE 4.1 Chemicals in the Workplace with Sufficient Evidence of Human Carcinogenicity*

Agent	Target Organs	Occupations
4-Aminobiphenyl	Urinary bladder	Dyestuff manufacturers; textile dyers; paint manufacturers
Arsenic and arsenic trioxide compounds	Skin, lung	Miners; smelters; insecticide and herbicide makers and sprayers; tanners; chemical workers; oil refining workers
Asbestos	Lung, larynx, GI tract	Miners; millers; textile, insulation, and shipyard workers; brake lining workers
Auramine (basic yellow 2)	Urinary bladder	Dyestuffs manufacturers and users; rubber workers; textile dyers; paint manufacturers
Benzene	Bone marrow	Explosives, rubber cement workers; distillers, dye users; painters; shoemakers; furniture finishers; glue and linoleum makers; petrochemical workers; styrene makers
Benzidine	Urinary bladder	Dyestuffs manufacturers and users; rubber workers; textile dyes; paint manufacturers
Bis(chloromethyl)ether and chloromethyl methyl ether	Lung	Ion-exchange resin makers; organic chemical synthesizers; polymer makers
Chromium and chromium compounds	Lung, nasal sinuses	Producers and processors; acetylene and aniline workers; bleachers; glass, pottery and linoleum workers; battery makers; anodizers; electroplaters; lithographers; photoengravers
Coke oven emissions	Lung, urinary tract	Coke oven workers
Hematite underground mining (iron ore, iron oxides)	Lung	Iron ore (hematite) miners; metal grinders and polishers; iron foundry workers
Isopropyl alcohol manufacture (strong acid process)	Paranasal sinuses, larynx, lung	Producers of isopropyl alcohol
Mustard gas	Lung	Mustard gas workers
β-Naphthylamine	Urinary bladder	Dyestuffs manufacturers and users; rubber workers; textile dyes; paint manufacturers
Nickel refining and nickel compounds	Lung, nasal passages, larynx	Nickel smelters; mixers and roasters; electrolysis workers

TABLE 4.1   Chemicals in the Workplace with Sufficient Evidence of Human Carcinogenicity* (continued)

Agent	Target Organs	Occupations
Soots, tars, and mineral oils (including creosote, shale and cutting oils)	Lung, urinary bladder, GI tract	Gashouse workers; asphalt, coal tar and pitch workers; coke-oven workers; miners; still cleaners; chimney sweeps; contact with lubricating, cooling, or fuel oils, or with paraffin, wax, coke or rubber fillers; retortmen textile weavers; diesel and jet testers
Vinyl chloride	Liver (angiosarcoma), brain, lung	Producers and polymer producers

* This listing excludes the pharmaceutical agents known to be carcinogenic.

production and consumption in the 1970s. About 2.5 million workers are estimated to have some exposure to asbestos. Cigarette smoking and asbestos act synergistically in increasing the risk of lung cancer. Asbestos exposure in the United States accounts for approximately 5% of lung cancer deaths and 1 to 2% of all cancer deaths in men. The principle form of lung cancer associated with exposure to asbestos is mesothelioma, which can occur in the pleura and in the peritoneum. Exposure to asbestos may also lead to an increased incidence of esophageal, gastric, and colorectal cancer.

Auramine (Basic Yellow 2)

Auramine is used industrially as dye or a dye intermediate for coloring textiles, leather, and paper. Approximately 3,000 workers are potentially exposed.

Benzene

Benzene is used extensively in industry as a solvent and a starting material or intermediate in the production of cyclic hydrocarbons. Three million workers are potentially exposed to benzene. Exposure to benzene is thought to be responsible for a higher than normal incidence of leukemia. Benzene is the only industrial chemical associated with the development of human cancer in the lymphoid tissues.

Benzidine

Benzidine is an aromatic amine used as an intermediate in the production of dyes and associated with the development of human bladder cancer. An estimated 2,200 workers are potentially exposed to benzidine. The annual production of benzidine has decreased significantly in the United States in the past several years.

Bis(chloromethyl)ether and Chloromethyl Methyl Ether

These compounds are used in the manufacture of plastics and ion exchange resins. Bis(chloromethyl)ether is associated with the development of small cell carcinomas of the lung. Chloromethyl methyl ether may also cause lung cancer, though the etiologic role is hard to prove because of contamination with bis(chloromethyl)ether.

Chromium and Chromium Compounds

Chromium is used in metal alloys, electroplating, magnetic tapes, pigment for paints, cement, rubber, and as an oxidant in the synthesis of organic chemicals. An increased risk of lung cancer was demonstrated in the chromate industry during World War II, and more recently an excess of lung cancer has been found in the chromate pigment industry. Hexavalent chromium compounds are thought to be responsible, but an exposure-response relationship has yet to be demonstrated. Currently, an estimated 2.5 million workers are exposed to chromium and its compounds.

Coke Oven Emissions

Coke oven emissions are a complex mixture resulting from coal combustion and distillation. They are composed of mostly polycyclic aromatic hydrocarbons and are associated with lung and urinary bladder cancers in humans. OSHA has estimated that 10,000 workers are exposed to these emissions.

Hematite (Iron Ore and Iron Oxides)

These compounds are associated with the development of lung cancer in man especially hematite miners. Exposure is through ingestion and inhalation of dusts which contain iron oxides, silica, and radon. The contribution of factors other than ferric oxide to increased lung cancer mortality in hematite miners is unknown.

Isopropyl Alcohol Manufacture

The strong sulphuric acid process used in the manufacture of isopropyl alcohol leaves a residue of isopropyl oils and diisopropylsulfate and diethylsulfate. It is not clear whether one or more of these substances is a respiratory tract carcinogen. The primary commercial uses of isopropyl alcohol are in acetone production and as a solvent or chemical intermediate. Approximately 141,000 workers may be potentially exposed to the isopropyl alcohol manufacturing process in the United States.

β-Naphthylamine

β-Naphthylamine is used principally as an intermediate in the manufacture of dyes and as an antioxidant in the rubber industry. There has been very little commercial production of β-naphthylamine in the United States during the past ten years.

Nickel Compounds

Nickel is used in electroplating, manufacturing of steel and other alloys, storage batteries, electric circuits, ceramics, petroleum refining, and oil hydrogenation. An estimated 710,000 workers are potentially exposed to nickel and related compounds.

Soots, Tars, and Mineral Oils (Including Creosote, Shale and Cutting Oils)

These materials result from fossil fuel processing technology and represent mixtures of aromatic hydrocarbons. The precise risk of exposure to these chemicals varies with the type of mixture and the type and route of exposure (i.e., inhalation, ingestion, or skin absorption).

Vinyl Chloride

The principal use of vinyl chloride is in the production of plastics, vinyl asbestos floor tiles, and packaging materials. More than 3.5 million workers are potentially exposed. The use of vinyl chloride as an aerosol propellant was banned by the Environmental Protection Agency and the Food and Drug Administration.

POSSIBLE CARCINOGENS IN OCCUPATIONAL SETTINGS

Table 4.2 is restricted to substances in the workplace for which the evidence of human carcinogenicity has been suggested on the basis of a casual observation rather than on the basis of a scientific observation.

By the mid-1960s, the discovery of new chemicals in the workplace for which there was persuasive evidence of human carcinogenicity was at an ebb. This changed in the early 1970s with the discovery of the carcinogenicity of vinyl chloride and bis(chloromethyl)ether. However, since 1973, several compounds have fallen under suspicion, but none has been recognized as a human carcinogen (Table 4.2). Cadmium is the only compound listed in this table for which there is persuasive evidence of human carcinogenicity (cancer of the prostate).

It appears that a stage has been reached where we no longer identify chemicals in the workplace that are clearly carcinogenic but also those whose danger is difficult to establish. Yet, it must be recognized that each of these "suspect" carcinogens should be considered as problematic, and exposure to them could be potentially harmful. Their borderline status probably results from the fact that they are weak carcinogens or because exposure to them is low.

TABLE 4.2   Chemicals in the Workplace for Which the Evidence of Human Carcinogenicity is Suspected but is Inadequate

Substance	Site	Occupations
Acrylonitrile	Colon, kidney, lung, urinary bladder	Acrylic resin, rubber and synthetic fiber makers; organic chemical synthesizers
Amitrole	Lung	Producers and herbicide users
Beryllium	Lung	Aerospace industry; nuclear reactor workers; beryllium refining and alloy workers; cathode ray tube makers; electronics workers; electric equipment workers
Cadmium	Prostate, kidney, lung	Alloy makers; battery makers; pesticide workers; textile printers; welders; zinc refiners; electroplaters
Carbon tetrachloride	Liver	Rubber, refrigerant and propellant makers; metal cleaners; chemists; solvent workers
Chlordane/Heptachlor	Central nervous system, hematopoietic tissues (leukemia)	Producers and pesticide users
Chloroprene	Liver (angiosarcoma), lung	Producers and polymer producers
Dimethylsulfate	Lung	Amine, dye and phenol derivative workers; organic chemical synthesizers
Epichlorohydrin	Hematopoietic tissues (leukemia), lung	Cellulose ether and solvent workers; epoxy resin, glycerol derivative, glycerophosphoric acid, glycidol derivative, paint, resin, and varnish makers
Ethylene oxide	Hematopoietic tissues (leukemia)	Detergent, disinfectant, ethylene glycol, polyglycol, and polyoxinane makers; sterilization room workers; organic chemical synthesizers
Hexachlorocyclohexane (BHCH/Lindane)	Hematopoietic tissues (leukemia)	Producers and pesticide users
Lead	Lung	Smelters; battery makers; dyers
Polychlorinated biphenyls	Skin	Dye, plasticizer, resin makers; rubber workers
Styrene	Lymphohematopoietic tissues (leukemia, lymphoma)	Petroleum refinery and solvent workers; organic chemical synthesizers; polyester resin laminators; polystyrene, resin and rubber workers

MECHANISMS OF CARCINOGENESIS

The most attractive hypothesis currently used to explain the initiation of carcinogenesis by chemicals involves the alteration of cellular DNA, with the ensuing mutations being passed on to succeeding cellular generations. Oncogenic chemicals are usually strongly electrophilic and form covalent bonds with nucleic acids. This results in mutations that may become permanent if not repaired by the cell. In several experimental systems, this initiating event, although necessary, is not sufficient to produce a tumor. Repetitive exposures to a tumor promoting agent must follow. Tumor promoters are agents that enhance cancer formation but by themselves are not carcinogenic. The mechanisms of tumor promotion are not fully known but include enhanced growth of initiated cells, alterations in cell-cell interactions, and induction of an inflammatory response in the target tissue. Several important occupational compounds have been identified as tumor promoters in rodent bioassays. These agents include solvents (e.g., trichloroethylene, carbon tetrachloride), pesticides (e.g., DDT, lindane), plasticizers (e.g., diethylhexylphthalate), and many peroxides (e.g., benzoyl peroxide). In addition, several DNA damaging, electrophilic carcinogens, such as 2-acetylaminofluorene, also have tumor promoting activity. Such agents are known as complete carcinogens, i.e., they can both initiate and promote carcinogenesis. Simultaneous exposure to more than one carcinogen may have an additive or even a synergistic effect on initiation and/or promotion. Most carcinogens must be metabolically activated before they become effective. Benzo(a)pyrene, for instance, is metabolized by cytochrome P-450 mixed function oxidases. One metabolite, a diol-epoxide, is highly mutagenic and can covalently bind to cellular DNA. The genetically determined levels of cytochrome P-450 enzymes, such as aryl hydrocarbon hydroxylase, may be important determinants of individual susceptibility to carcinogens. Nutritional factors may also be important, as evidenced by a reduced incidence of tumors in animals receiving vitamin A or other inhibitors of carcinogenesis.

The induction of mutations provides a possible explanation for the carcinogenic activity of many chemicals. Several carcinogens are not mutagenic, however. Asbestos is one such agent; the mechanism by which it produces cancer in humans is not understood. Several theories for the oncogenic potential of asbestos have been proposed; the physical characteristics of the fiber appear to be of paramount importance in asbestos-induced mesothelioma. In addition, immunologic abnormalities have been reported in workers with asbestosis and may predispose them to lung cancer.

CANCER EPIDEMIOLOGY

The purpose of cancer epidemiology is to identify situations in which there is an increased risk of cancer. Even when it is not feasible to accurately identify a carcinogen or its mechanism of action, the demonstration of consistent associations between certain occupations and cancer is often sufficient to allow the disease to be prevented. In epidemiology, the people being studied have generally been exposed to the prevailing levels of the suspected chemicals so that extrapolation from greatly different dosages and from other species is avoided. Epidemiologists are sometimes able to study large populations so that even small effects on health may be detected. However, the inability to control modifying or confounding factors and the possibility of bias in selecting study populations can result in difficult interpretations. In addition, the long latent period of cancer often means that an important risk can be identified only after extensive exposure to the agent in question and other agents has occurred, further confounding interpretations.

Three basic strategies are employed in epidemiological research. *Cross-sectional surveys* on the prevalence of disease are useful for defining the effects of a known or suspected risk in a group of workers. However, these are studies of survivors, and although they can reveal association between exposure and disease, it is very difficult to assess causality. In *cohort studies*, a defined population is observed over time for the development of disease; this approach provides the most reliable information on an exposure/disease relationship. However, long-term observation is very difficult and expensive, and occupational cancers tend to occur so infrequently that the number of new cases is too small to be useful. Consequently, cancer epidemiology is usually dependent on *case-reference studies*, in which the extent of exposure in subjects with or without a specific cancer is compared. This has been the most useful approach to identify putative human carcinogens in occupational settings.

PREVENTION OF OCCUPATIONAL CANCER

Prevention of occupational cancer entails the identification of carcinogenic substances in the workplace and the reduction of human exposure to these substances. Animal bioassays are used to identify carcinogens but are very expensive and time consuming. To test a chemical for carcinogenicity in animals, it must be administered to two different species during their lifetime; in the United States the cost of doing this is nearly one million dollars over 3 years for each substance. Moreover, the results are not readily applicable to the human situation, primarily because of interspecies differences in carcinogenic susceptibility. The fungal toxin, aflatoxin B_1, for example, is a very potent liver carcinogen in the rat but is only weakly carcinogenic in the mouse. One might expect the differences between rodents and humans to be even greater. Further difficulties in interpretation may arise because of the different routes of entry used in animal studies, the high dosages employed in experimental situations, and the lack of statistical power involved when relatively small numbers of animals are studied.

Since 1975 the role of short-term or *in vitro* tests for identifying environmental carcinogens has been expanding. The advantages of these tests are their rapidity and relatively low cost. The prototype was the *Salmonella* mutagenesis test developed by Ames and co-workers, in which mutations are detected in bacterial strains that have been exposed to suspected carcinogens. Similar tests now use fungi, cell cultures from different organisms, or whole animals such as Drosophila. The endpoints can be chromosomal damage, sister chromatid exchanges, unscheduled DNA synthesis, or cell transformation.

All of these tests are based on the hypothesis that agents capable of inducing DNA damage can cause cancer if they gain access to target cells in the body. While this seems plausible in light of current scientific knowledge, not all mutagens are carcinogenic. Furthermore, not all important causes of cancer (tumor promoters and asbestos for example) are mutagenic. Most *in vitro* tests are thus unlikely to detect nonmutagenic carcinogens. However, many tumor promoters have the ability to inhibit cell-cell communication, through gap junctions. This effect may be exploited in the future for the rapid identification of promoting agents. However, cell-cell communication assays also have short-comings and, therefore, a battery of short-term tests may be needed to detect all potential carcinogens.

Biochemical epidemiology is another approach that is proving useful for identifying individuals at high risk for cancer through exposure to carcinogens in occupational settings. Laboratory investigations with cultured human tissues and cells have shown that humans vary markedly with respect to the extent of genetic damage they receive through exposure to environmental carcinogens. The availability of sensitive techniques such as radioimmunoassay, enzyme-linked immunoassay, synchronous scanning fluorescence, and ^{32}P-postlabeling to detect and quantitate chemical-DNA adducts easily obtainable in human cells (e.g., blood lymphocytes), or immunoassays to detect chemical-hemoglobin adducts renders it possible to monitor exposure of humans to chemicals in specific environmental settings. Long-term prospective studies using these methodologies may eventually permit the identification of individuals at high risk to the development of cancer through exposure to specific industrial chem-icals. High-risk individuals could then be cautioned against working in occu-pational settings in which exposure to the chemical(s) is extensive.

FINAL COMMENTS

Although probably fewer than 5% of cancers can be attributed to occupation, the social impact of malignant diseases related to the workplace is great. The public has great concern and often alarm concerning occupational exposure to suspected carcinogens.

Since the initial detection of occupational hazards is usually dependent upon clinical observations, the physician has a major role to play in identifying carcinogens. The diagnosis of an unusual type of tumor or of a common tumor in an unusually young patient or one who is without recognized risk factors is an important clue to the presence of an environmental carcinogen. The clustering of cases in particular industrial settings can be discerned by taking an occupational history from every patient who has cancer. The principal methods for reducing the risk of occupational cancer include the identification of substances in the workplace that induce cancer and the reduction of exposure of workers to these substances.

REFERENCES

Hall, S. K., Chemical structure and carcinogenesis, *Chemistry in Canada*, 29:18–23, 1977.

Harris, C. C., Human tissues and cells in carcinogenesis research, *Cancer Research*, 47:1–11, 1987.

Sax, N. I., *Cancer Causing Chemicals*, Van Nostrand Reinhold, New York, 1981.

Schottenfeld, D., Chronic disease in the workplace and environment: Cancer, *Archives of Environmental Health*, 39:150–157, 1984.

Spivy, A. M., *Occupational Carcinogens in Perspective*, Elsevier Science, Oxford, England, 1988.

Stellman, J. M. and Stellman, S. D., Cancer and the workplace, *CA Cancer Journal*, 46:70–92, 1996.

5

Chemical Exposure and Hypersensitivity

Stephen K. Hall

CONTENTS

INTRODUCTION

The massive increase in our environmental exposure to chemicals, both synthetic and natural, has altered our bodily levels of these substances. It is now all but impossible to find an individual who does not have a detectable level of synthetic chemicals in his or her body. Environmental concentrations of natural chemicals such as ozone and oxides of nitrogen are many orders of magnitude higher than in the past. Humans have many biochemical scavenger systems that protect them from damage caused by chemically altered cells and proteins. However, since we are now exposed to much higher concentrations of natural chemicals, as well as massive amounts of synthetic chemicals to which our ancestors were never exposed, it is obvious that our protective resources are taxed to a much greater extent than were those of our ancestors.

Occupational health and safety professionals frequently encounter one or more individuals in a plant with a remarkable ability to detect chemical leaks, the escape of any solvent vapor or any toxic fumes, or breakdowns in the exhaust system. These individuals can detect the presence of chemicals in the air long before anyone else notices them. In general, these individuals have something in common. They have increased susceptibility to being overcome by solvents or chemical odors. They feel light-headed and queasy, and experience headache, nasal congestion, and throat irritation following the least exposure to a variety of chemical odors. They are barely tolerated by their coworkers and are sometimes regarded as being crazy. These individuals in fact suffer from chemical hypersensitivity syndrome.

Chemical hypersensitivity syndrome (CHS) is a chronic (i.e., continuous for more than three months), multi-system disorder, usually involving symptoms of the central nervous system and at least one other system. Affected individuals react adversely to some chemicals and to environmental agents, singly or in combination, at levels generally tolerated by the majority of the population. Affected individuals have varying degrees of morbidity, from mild discomfort to total disability. Upon physical examination, the individual is normally free from any abnormal, objective findings. No single clinical laboratory test is consistently altered. Improvement is associated with avoidance of suspected agents and symptoms recur with re-exposure.

Chemical hypersensitivity syndrome is the name most frequently used in industry. A variety of other names have also appeared in the literature and they are recorded in Table 5.1.

TABLE 5.1   Environmental Illness: Synonyms

Allergic toxemia
Cerebral allergy
Chemical hypersensitivity syndrome
Chemically induced immune dysregulation
Clinical ecology
Complex allergy
Ecologic illness
Environmental hypersensitivity disorder
Multiple chemical sensitivity
Total allergy syndrome
Twentieth century disease

INDIVIDUALS WITH CHEMICAL HYPERSENSITIVITY SYNDROME

Much of our population is reacting adversely to the chemicals around us. In recent years, increasing attention has been directed toward the assessment of the impact of occupational and environment exposures on human health. The concern has generated increasing public involvement. For example, the Ministry of Health of the Province of Ontario, Canada, commissioned a study on Environmental Hypersensitivity Disorders which was completed in 1985. The Committee placed newspaper advertisements in 16 Ontario newspapers soliciting testimonials from individuals and organizations concerned about chemical hypersensitivity. Data were received from 130 individuals, representing

119 families. Seventy-five percent of the respondents were female, in their 30s, and of middle class socioeconomic status. The median number of years of symptomatology was eight. Primary systems reported as affected were the central nervous system (77%), respiratory system (45%), and gastrointestinal system (44%).

In an independent study in the United States, a review of 100 consecutive cases admitted to the environmental illness unit in Canelton, Texas, confirmed many of the Canadian observations. The study population was predominantly female (77%), with a median age of 40. Of the group, 54% were highly educated, with four or more years of college. Most (56%) first experienced symptoms before they were 30 years old. Forty-three percent reported exposure to chemicals on their job as responsible for their conditions.

There are several factors that may provide insight into the prevalence of CHS. First it is important to acknowledge the increased presence of a variety of chemical substances in our environment of which the acute and chronic effects have not fully been evaluated. Recent evidence suggests a portion of the population is sensitive to quite low levels of environmental contaminants.

This phenomenon has increased in the years following the institution of energy-saving measures, including "sealing" of buildings from natural ventilation. While some individuals have developed CHS following tight building syndromes, the overwhelming majority have not. The incidents are instructive, however, in that they provide evidence of the irritant effects of low-level exposures, as well as the variability of individual reactivity.

DIAGNOSTIC APPROACHES

Chemical hypersensitivity syndrome is an acquired disorder characterized by recurrent symptoms, referable to multiple organ systems, occurring in response to demonstrable exposure to many chemically unrelated compounds at doses far below those established in the general population to cause harmful effects. No single widely accepted test of physiologic foundation can be shown to correlate with symptoms.

Although many of the victims of CHS feel too sick to work, assessment of impairments, disability, and attribution or cause is hindered by the current lack of clear diagnostic criteria, frequent lack of chemical exposure data, and inadequate tools to evaluate the extent of impairment. This situation often leads to frustration on the part of the physician, resulting in summary judgments that the individual's problems are psychosomatic in origin.

The victims are often searching sincerely for assistance, which may be medical, psychosocial, or economic in nature. Often victims are subject to simultaneous stress and anxiety due to the perplexing nature of their symptoms, perceived inability to continue to work in an environment that apparently exacerbates their condition, and social pressures to continue to tolerate what he perceives as severe but not externally evident symptoms such as headache, difficulty concentrating, nausea, fatigue, somatic pain, or dizziness. At the same time they face economic pressures from threats to their income and costs of medical care. The usually safe refuges such as home or office may be sources of potential aggravation of the symptoms, leading to dramatic changes in lifestyle, habitat, and social contact based on efforts to avoid exposures to potential chemical toxicants.

Diagnosis of the victims with CHS is usually possible, by a combination of a detailed history, physical examination, and a few specific biochemical tests. A history of allergy, atopy or asthma should also be sought. A detailed occupational and environmental history is, of course, crucial for assessing potential chemical exposures. Industrial hygiene data, if available, may be useful in documenting the extent of exposure. Often, however, the recurrent syndrome is provoked by exposure levels that are low and well within the current occupational exposure limits.

Physical examination should be performed with particular attention to the skin, head, ears, eyes, nose, throat, respiratory tract, gastrointestinal tract, and the central and peripheral nervous systems. Often, however, the physical examination is completely normal. The history and physical examination is frequently sufficient to exclude as reasonable possibilities an allergic disorder or classic occupational disease. Neuropsychological testing, if required, should be performed by an experienced clinical neuropsychologist. It is most useful in the evaluation of individuals with a history of central nervous system symptoms or findings suggestive of cognitive or motor impairment on mental status examination.

The concept of chemical hypersensitivity as a disease entity in which the individual experiences numerous symptoms from numerous chemicals caused by a disturbance of the immune systems lacks a scientific foundation. Published reports of such cases are anecdotal and without proper controls. There is no convincing evidence for any immunologic abnormality in the reported cases. Immunologic diagnosis, treatment, and theoretical concepts are not consistent with current immunologic knowledge and theory.

Individuals with CHS show numerous physiological and biochemical abnormalities and are generally sicker than a control group of allergic individuals. Associated with CHS are hypothyroidism, autoimmune thyroiditis, mitral valve prolapse, specific abnormalities of amino acid and essential fatty acid metabolism and diminished activity of erythrocyte activity of superoxide dismutase and erythrocyte glutathione peroxidase activity. Equally prevalent among CHS individuals and control groups of allergic individuals are deficiencies of magnesium and vitamin B_6.

The typical symptoms experienced by individuals with CHS (e.g., nasal congestion, throat or eye irritation, hoarseness, nausea, chest pain, palpitations, dyspnea, choking or smothering sensations, tremulousness, headache, lightheadedness, dizziness, faintness, fatigue, feelings of unreality, etc.) are relatively ubiquitous in the general population. Many of these are also the known physiologic concomitants of anxiety and fear. Individuals vary in their response to commonly occurring symptoms. Most people experience these symptoms only transiently and then dampen their response; they dismiss the symptoms as insignificant and divert their attention to other matters. Individuals with CHS, however, become increasingly anxious and aroused and symptomatic following the onset of symptoms; they tend to amplify rather than dampen their symptoms.

In general, the following major diagnostic features can be distinguished in individuals with CHS:

1. The disorder is acquired in relation to some documentable environmental exposure;
2. Symptoms involve more than one organ system;
3. Symptoms recur and abate in response to predictable stimuli;
4. Symptoms are elicited by exposures to chemicals of diverse structural classes and toxicologic modes of action;
5. Symptoms are elicited by exposures that are demonstrable;
6. Exposures that elicit symptoms must be very low, by which we mean many standard deviation from "average" exposures known to cause adverse human response.

MANAGEMENT OF THE WORKER WITH CHEMICAL HYPERSENSITIVITY

The difficult task of managing individuals with CHS requires a multi-disciplinary effort involving the industrial hygienist, social worker, psychologist, occupational therapist, occupational health nurse, occupational medical physician, and other involved professionals. Important objectives of management include complete review of the occupational history, search for additional exposure information, appropriate choice of laboratory tests, emphasis on "well" behavior, health education of the individual and supportive therapy. Because of the chronic nature of CHS and the difficulty many individuals have in coping with the illness, pharmacologic and psychologic interventions should be considered.

INDUSTRIAL HYGIENE INTERVENTION

According to the American Conference of Governmental Industrial Hygienists, the charge of traditional industrial hygiene is to protect "nearly all workers." Guidelines are specified "to represent conditions in which it is believed that nearly all workers may be repeatedly exposed day after day without adverse effect." However, it is acknowledged that some workers may experience discomfort at concentrations of some substances below the Threshold Limit Values (TLVs) and even fewer may be affected more seriously because of a pre-existing condition or by development of an occupational illness. Obviously, workers with CHS are certainly among those who would fall into the group of individuals at the extreme low end of the dose-response curve.

Industrial hygiene intervention should begin with a complete occupational history, starting with the current or most recent job and going back as far as possible to obtain the entire history of the individual's working life. Find out what the individual was exposed to and to what concentration in each of the job environments. Particular attention should be focused on the time period of onset of symptoms, and the chemicals or materials handled.

Similarly, information about non-work environments must be obtained. The focus here is on the changes that may have occurred prior to or during the beginning of the illness. Pest extermination, home insulation or renovation may be significant.

Partly due to training and partly due to regulatory requirements, industrial hygienists tend to approach the environment by attempting to qualify its contaminants. Usually this is done without detailed knowledge of the clinical complaints that may have prompted the investigation nor any epidemiologic information suggesting a relationship between the workplace and complaints. In evaluating the environments of individuals with CHS this approach may be misleading, since it almost invariably leads to the premature conclusion that exposures are "too low" to be causing effects. This circumvents the real issue and may lead the individual and management into polarized, intransigent views of the real problem.

While quantifying levels of particular contaminants is often unrewarding, extensive assessment of general as well as local ventilation is frequently of major importance. Bringing the ventilation system up to standard eliminates indoor air pollution problems in the vast majority of workplaces. The system must be inspected to make sure it has been properly maintained, that the fans are functioning as they should, and that fresh air dampers are well placed away from contamination sources and sufficiently open. Then the amount of fresh air actually being delivered must be calculated.

The American Society of Heating, Refrigerating and Air-Conditioning Engineers (ASHRAE) has developed guidelines suggesting a minimum of 20 cubic feet per minute (cfm) of fresh air per person, to maintain a carbon dioxide concentration of less than 1,000 parts per million (ppm). It has been recommended that new buildings be ventilated with 100% fresh air for the first 6 months, and that recirculated air be limited to a maximum of 50% for the next 1 to 2 years because of off-gassing of new building materials.

PSYCHOLOGICAL INTERVENTION

Treatment of CHS must incorporate multiple types of help, all directed toward supplying what these individuals require. As in all practice in the medical setting, the overall function of the social worker is to enable the individual with CHS to make use of what the physician has to offer by supporting the individual's capacity to cope with the social and emotional impact of the individual's illness.

Reports of exposure leading to the development of CHS typically involve feelings of inability to maintain control over the circumstances surrounding the exposure and their reactions to the exposure. Intervention at the early stage with the CHS individual as the problem requiring immediate attention. The emphasis should not be on the development of understanding by the individual but rather on reinforcing strengths through guidance, release of tension and through reassurance. Group treatment may be suggested if the individual is felt to be appropriate for this modality.

Maintaining or returning a sense of control to these individuals is important in treatment planning. Education provides a potential means to promote well being and behavior and to prevent further illness. It also provides tools for the individual to control his environment rather than encourage continued illness, disability and hopelessness. Education does not necessarily lead to a change in behavior or attitudes, but the social worker can transmit his knowledge to the individual in such a manner that it will optimize the individual's understanding of the known versus implausible health effects expected as a consequence of their real versus imagined exposures.

BIOLOGICAL INTERVENTION

Two key questions should frame the biological intervention plan for the individual with CHS. What is the individual sensitive to and why is the individual so sensitive? The answer to the first question is used to design a program of avoidance that will reduce symptoms. Investigating the second point is both challenging and potentially rewarding.

A careful medical, environmental, and occupational history will elicit the onset and recurrent precipitation of symptoms by certain exposures. In contrast to the picture of a person with a specific sensitivity, individuals with CHS reveal a broad spectrum of sensitivities to offensive agents. Despite controversy about validity, reliability, and significance, provocative challenge, antibody testing, and direct biological assay for specific substances have been used to direct attention to groups of agents, organic solvents, pesticides, or gaseous irritants.

Two key hypotheses are relevant to an avoidance program. The first is the concept of "total environmental load" (TEL). This idea corresponds to the traditional industrial hygiene concept of "Threshold Limit Values for Mixtures". The TEL expands the industrial hygiene precept that knowledge of the sum of all toxic exposures to an individual is key to predicting health effects. In theory reducing the TEL will decrease an individual's sensitivities to any given substance. The second hypothesis stems from the clinical observation that chronic symptoms in some individuals may be the result of imperfect physiologic adaptation to on-going low-level chemical exposures. In other words, the temporal relationship between symptoms and exposures is masked until "total chemical avoidance" is achieved. Total avoidance unmasks exposure/symptoms relationship. In effect, the individual becomes more sensitive after brief total avoidance. Accordingly, prolonged avoidance will result in decreasing sensitivity and renewed tolerance.

A general biological intervention theory of CHS has been proposed. It suggests that sensitivity occurs as normal antioxidant defenses are overwhelmed either by exposure to exogenous oxidizing chemicals or through infection that stimulates phagocytic production of free radicals. It also notes that numerous representatives of the chemical classes that cause chemical sensitivity reactions (e.g., aromatic hydrocarbons, chlorinated hydrocarbons, or the photochemical oxidants) exist either as free radicals in the aqueous tissue environment or are capable of being metabolized to free-radical intermediates.

The individual's nutritional status also plays an important role in determining his or her response to certain chemical exposures. A wide range of nutritional disturbances in individuals with CHS has been documented. Whether these problems are the result of the illness or the behavior associated with CHS or predate the syndrome is unknown. In either case, nutritional disorders may compound CHS. Use of certain nutrients in pharmacologic doses may improve an individual's tolerance to chemicals and overall well being.

The initial emphasis in nutritional intervention is on insuring a balanced high-fiber diet that minimizes intake of synthetic chemicals and on replacing specific nutrients found deficient in the biochemical examination. Dietary fats are manipulated to decrease total fat intake to about 30% of the calories, with an equal division between saturated, cold-pressed, nonhydrogenated, polyunsaturated, and monounsaturated fats. If fat is used in cooking, saturated and mono-unsaturated fats are recommended. This plan effectively decreases the possible oxidant load from fat and minimizes possible fatty acid metabolic disturbances resulting from a diet high in saturated fats.

HEALTH CARE INTERVENTION

Treatment of individuals with CHS often receives less attention by clinicians in health care facilities than diagnostic evaluation. Whether this is due to inadequate knowledge of how to treat such individuals or to some frustration because these individuals are "difficult," the lack of treatment as a priority is an error that prolongs the individual's symptoms and impairments.

The treatment of individuals with CHS must always begin with a thorough review of the history, physical examination and diagnostic tests already performed. The staff of the health care facilities must incorporate into their interpretation of the problem an understanding that these individuals have often already had multiple diagnoses with multiple diagnostic tests, both necessary and unnecessary, and that the individuals' lives may be in considerable turmoil. Furthermore, the individual may already have alienated family, friends, personal physicians, and employers, and had major changes in income, lifestyle, and attitudes toward work.

Appropriate management of the individual with CHS requires a full understanding of the individual's interaction with his social environment. It must be emphasized that the entire clinic staff should have as complete an understanding of the social interactions of the individual as possible. The staff must have skill in providing support in the individual's relationships with family, employers, coworkers, legal advocates, other health providers, insurance carriers, as well as many others. Because the symptoms are nonspecific, the physical examination unremarkable, and the laboratory tests are inconclusive, the diagnosis may be unclear and the specific etiology of the condition may be undefined. As a result, the individual's complaints are often viewed with skepticism by some clinic staff. Lack of sympathy and negative reactions by clinic staff often complicate the individual's life. The clinic staff must therefore be alert to these problems and provide appropriate support to minimize these problems.

If the individual's symptoms are diagnosed to be caused by chemical exposures in the workplace, for example, organic solvent vapors, it is expected that ongoing exposures, even to low levels, may continue to cause symptoms and prolong the illness. Even when an organic basis for the individual's complaints is not identified, removal from exposure to the alleged offending chemicals may be advisable.

The individual's interaction with the employer is often especially troublesome. The clinic staff must take great care to communicate appropriately with the employer and accurately advise the individual regarding the work-relatedness of the condition and the recommended work restrictions. Often the individual already has made a decision not to work by the time of the initial visit to the clinic. If the individual stops working without adequate justification, without a clear statement of the necessity for removal from exposure, without a description of the likely consequences of re-exposure, and without a definition of the objectives, the employer is more likely to be unsympathetic. The clinic must take steps to communicate clearly with the employer. If the individual came to the clinic on his or her own, or was referred by someone other than his or her employer, appropriate consent must be given by the individual to communicate with the employer. If a work-related disease is diagnosed, it should be clearly communicated. Any work restrictions that may be advisable, regardless of etiology, should be indicated, as well as the duration of the recommendation. The anticipated consequences of continued exposure or noncompliance with the recommendations should be specified.

Long ago, psychiatry recognized that it had no treatments that would resolve the problems and symptoms of many of its individuals, but that these sometimes desperate individuals needed human contact, advice, an opportunity to ventilate, and reassurance. Clinicians must provide supportive treatment for CHS individuals. Such treatment should consist of acknowledging that the individual's symptoms are very real and very frightening for the individual. In order to reassure the individual, thorough physical examination and laboratory studies should be conducted. The clinician should discuss the results of testing with the individual in great detail and the individual should be told what the truth is. The clinician should accept the individual's condition as chronic and should help the individual accept that condition.

REFERENCES

Cullen, M. R. (ed), *Workers with Multiple Chemical Sensitivities*, Hanley and Belfus, Inc., Philadelphia, 1987.

Hall, S. K., The worker with chemical hypersensitivity syndrome, *Pollution Engineering*, 21(2):76–79, 1989.

Kimber, I. and Maurer, T., *Toxicology of Contact Hypersensitivity*, Taylor & Francis, Philadelphia, 1996.

Report of the Ad Hoc Committee on Environmental Hypersensitivity Disorders, Office of the Minister of Health, Toronto, Ontario, Canada, 1985.

Rest, K., *Proceedings of Multiple Chemical Sensitivity Workshop*, Princeton Scientific, Princeton, 1992.

Part II: Toxic Responses

6

Toxic Responses of the Lung

Stephen K. Hall

CONTENTS

1-56670-239-9/97/$0.00+$.50
© 1997 by CRC Press, Inc.

INTRODUCTION: THE RESPIRATORY SYSTEM

The respiratory system consists of all the organs of the body that contribute to normal respiration. Strictly speaking, it includes the nose, mouth, upper throat, larynx, trachea, and bronchi, which are all air passages. It also includes the lungs, where oxygen is passed into the blood and carbon dioxide is given off. The lungs represent a quick and direct avenue of entry for toxic materials into the body because of its intimate association with the circulatory system. Finally, the respiratory system (Figure 6.1) includes the diaphragm and the muscles of the chest, which permit normal respiratory movements.

THE RESPIRATORY SYSTEM

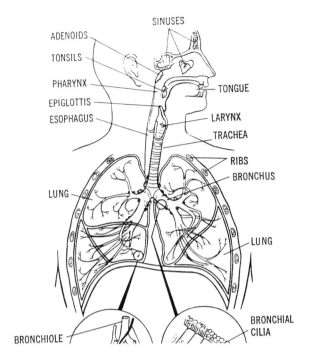

FIGURE 6.1   The respiratory system.

The Nose

The nose consists of an external and internal portion. The upper part of the external portion is held in a fixed position by the supporting nasal bones that form the bridge of the nose. The lower portion is movable because of its pliable framework of fibrous tissue, cartilage, and skin. The internal portion of the nose lies within the skull between the base of the cranium and the roof of the mouth, and is in front of the nasopharynx. The nasal septum is a narrow partition that divides the nose into right and left nasal cavities. The nasal cavities open into the nasopharynx. The vestibule of each cavity is the dilated portion just inside the nostril. Toward the front, the lining of the vestibule is lined with skin and represents a ring of coarse hairs which serve to trap dust particles. Toward the rear, the lining of the vestibule changes from skin to a highly vascular ciliated mucous membrane, called the nasal mucosa, which lines the rest of the nasal cavity.

Air enters through the nostrils, passes through a web of nasal hairs, and flows posteriorly toward the nasopharynx. The air is warmed and moistened in its passage and partially depleted of particles. Some particles are removed by impaction on the nasal hairs and at bends in the air path, and others by sedimentation. In mouth breathing, some particles are deposited, primarily by impaction, in the oral cavity and at the back of the throat. These particles are rapidly passed to the esophagus by swallowing.

Near the middle of the nasal cavity, and on both sides of the septum, are a series of scroll-like bones called the turbinate. The surfaces of the turbinate, like the rest of the interior walls of the nose, are covered with mucous membranes. The membranes secrete a fluid called mucus, which serves as a trap for bacteria and dust in the air. In addition to the mucus, the membrane is coated with cilia, or hairlike filaments, that move in coordinated waves to propel mucus and trapped particles toward the nostrils. The millions of cilia lining the nasal cavity help the mucus clean the incoming air. When breathing through the mouth, the protective benefits of the cilia and mucus are lost.

The Pharynx

From the nasal cavity, air moves into the pharynx. The pharynx is a tubular passageway attached to the base of the skull and extending downward behind the nasal cavity, the mouth, and the larynx to continue as the esophagus. Its walls are composed of skeletal muscle, and the lining consists of mucous membrane.

The nasopharynx is the superior portion of the pharyngeal cavity. It lies behind the nasal cavities and above the level of the soft palate. The function of this portion of the pharynx is purely respiratory, and its ciliated mucosal lining is continuous with that of the nasal cavities.

At the bottom of the throat are two passageways: the esophagus behind and the trachea in front. Food and liquids entering the pharynx pass into the esophagus. Air and other gases enter the trachea to go to the lung. Guarding the opening of the trachea is a thin structure called the epiglottis. This structure helps food glide from the mouth to the esophagus.

The Larynx

The larynx serves as a passageway for air between the pharynx and the trachea. It is lined with ciliated mucous membrane and the cilia move particles upward to the pharynx. The unique structure of the larynx enables it to function somewhat like a valve on guard duty at the entrance to the trachea, controlling air flow and preventing anything but air from entering the lower air passages.

The Trachea

The trachea extends from the bottom of the larynx through the neck and into the chest cavity. At its lower end it divides into two tubes, the right and left bronchi. Rings of cartilage hold the trachea and bronchi open. The wall of the trachea is lined with mucous membrane, and there are many hairlike cilia fanning upward toward the throat, moving dust particles that have been caught in the mucous membrane, thus preventing them from reaching the lung.

The Bronchi

The trachea divides into the right and left bronchi. Each bronchus enters the lung of its own side. The right bronchus is wider and shorter than the left. Its direction is almost identical with that of the trachea. This is why most aspirated material enters the right lung. Each bronchus divides and subdivides into continually smaller, finer, and more numerous tubes, something similar to the branches of a tree. Accordingly, the whole structure is sometimes called the bronchial tree. In the larger branches there also is stiffening by rings of cartilage, but as the branches get smaller, the cartilage becomes reduced to small plates and finally disappears.

The Lungs

There are two lungs, one on each side of the thoracic cage. The lungs are suspended within the thoracic cage by the trachea, by the arteries and veins running to and from the heart, and by pulmonary ligaments. The lungs extend from the collarbone to the diaphragm. Taken together, they fill almost all of the thoracic cavity. The two lungs are not mirror images of each other. The right lung is slightly larger of the two and partially divided into three lobes. The left lung is divided into two lobes.

The lungs are covered by a double membrane. One, the pleural membrane, lies over the lungs. The other lines the chest cavity. They are separated by a thin layer of fluid that prevents the two membranes from rubbing against each other.

Within a lung, the bronchi divide and subdivide, becoming smaller and smaller, until the branches reach a very fine size, which are bronchiolus. The respiratory bronchiolus lead into several ducts. Each duct ends in a cluster of air sacs called alveoli.

The respiratory tract, with its successive branches and tortuous passageways, is a highly efficient dust collector. Essentially all particles entering the respiratory tract larger than 4 or 5 μm are deposited in it. About half of those of 1-μm size appear to be deposited and the other half exhaled. Particles greater than 2.5 to 3 μm in size are deposited, for the most part, in the upper respiratory system, i.e., the nasal cavity, the trachea, the bronchi, and other air passages; whereas particles 2 μm in size are deposited about equally in the upper respiratory system and in the alveolar air spaces. Particles about 1 μm in size are deposited more efficiently in the alveolar spaces than elsewhere; essentially none are collected in the upper respiratory system.

EFFECTS OF SOME INHALED OCCUPATIONAL AGENTS

Some inhaled chemical agents are hazardous primarily to the respiratory system while others also affect distant organs and tissues. Examples of the former are ammonia and chlorine while examples of the latter are nitrogen and ozone. How extensively the respiratory system is involved, particularly following acute or accidental exposure, is determined largely by the concentration of the agent and duration of exposure. Other factors that may modify the individual's response include pre-existing heart or lung disease, prior long-term exposure to the same agent, level of activity during exposure, and the individual's age.

The signs and symptoms of mild exposure to irritant gases that are relatively soluble in aqueous solutions (e.g., ammonia, chlorine, and sulfur dioxide) are likely to be confined to the upper respiratory system. In response, an individual may experience one or more of the following: sneezing, nasal catarrh, unpleasant smell or taste, soreness of the throat, smarting of the eyes, and lacrimation. More intense exposure extends the involvement to the central airways of the tracheobronchial tree. Cough, sputum, pain, or constriction of the chest, and in the event of bronchospasm, shortness of breath and wheezing, may be present. The most intense exposures damage the alveolar-capillary membrane of the lung. The consequence is edema. Depending on the amount of edema that forms, an individual may suffer extreme shortness of breath, dusky discoloration of the mucus membranes and cyanosis of the nail beds, blood-tinged sputum mixed with foam, and collapse.

Occupational agents with relatively low solubility in aqueous solution tend to shift their primary effect to the periphery of the respiratory system. Thus, nitrogen dioxide and ozone are notable for the bronchiolar and parenchymal injury that they produce at relatively low concentrations.

The effect of an irritant or toxic agent contained in an inhaled particle is intimately related to the aerodynamic behavior of the particle, since aerodynamic behavior is a determinant of where and how much deposition occurs within the respiratory system. Cadmium, mercury, and sulfuric acid are examples of occupational agents that are inhaled in particulate form.

The respiratory system has several means of clearing itself of foreign particles. Most solid particles that deposit in the alveolar region are engulfed by macrophages. These are mobile cells that transfer the material to nearby terminal airways. A small, variable fraction of these particles may pierce the alveolar lining and either imbed in fixed tissues or be removed through lymphatics or blood vessels. The mucociliary system, beginning with the terminal bronchioles, carries particles from the nasal passages and from the lower airways toward the throat. The particles are then swallowed or expectorated. Cough is effective in clearing the central airways.

Ammonia

Ammonia is a colorless, strongly alkaline, extremely soluble gas with an easily recognized odor. It is used in the production of fertilizers and in the manufacture of nitric acid, soda, synthetic urea, synthetic fibers, dyes, and plastics. Other sources of occupational exposure include the silvering of mirrors, glue making, and tanning of leather. Contact with ammonia is intensely irritating to the mucous membranes, eyes, and skin. Eye symptoms range from lacrimation and edema, to a rise of intraocular pressure, ulceration, and blindness.

There may be corrosive burns of the skin or blister formation. Bronchitis or pneumonia may follow a severe exposure. Massive accidental exposure to ammonia can be rapidly fatal.

Arsine

Arsine is a colorless, very highly toxic gas with a faint, garlic-like odor. Although arsine as such is not used in industry, it is apt to be encountered wherever arsenic, even in scrap metals, becomes moist under reducing conditions with free hydrogen. In addition, arsine is involved in the smelting and refining of various arsenic-containing ores. Upon inhalation, arsine damages lungs and passes into the bloodstream where it causes hemolytic destruction of the red blood cells. Invariably, the sign observed in arsine poisoning is hemoglobinemia, appearing with discoloration of the urine up to port wine hue. Jaundice sets in on the second or third day and may be intense. Coincident with these effects is a severe hemolytic-type anemia. Severe renal damage may occur with oliguria or complete suppression of urinary function, leading to uremia and death. Where death does not occur, recovery is prolonged.

Chlorine

Chlorine is the most abundant halogen and among the most reactive of all elements. It is a yellowish-green gas with a pungent, irritating odor that is readily detectable in a concentration of 3.5 ppm. Chlorine has use as a disinfectant but is more widely used as a bleach by many industries. It is also used in refining petroleum and the preparation of many chloride compounds. Chlorine is a powerful irritant of the mucous membranes of the eyes and upper respiratory tract. Upon absorption into tissue fluids, chlorine undergoes a series of reactions to produce hydrochloric acid, hypochlorous acid and nascent oxygen. Each of these chemicals damage biologic tissue. The consequences of accidental overexposure to chlorine gas are well documented. In addition to acute inflammation of the eyes, entire respiratory system and skin, the teeth may be damaged or discolored. Death may be caused by asphyxia from laryngospasm or massive pulmonary edema.

Hydrogen Sulfide

Hydrogen sulfide is a colorless, flammable, and explosive gas with a powerful nauseating odor. The smell is generally perceptible at 0.77 ppm; at 4.6 ppm it is quite noticeable; while at 27 ppm, it is strong and unpleasant. Hydrogen sulfide is not used directly in industry but is formed frequently as a by-product of certain processes. It is found in the sulfur dye industries, in tanneries, in the production of carbon disulfide, and in the heating of some rubbers containing sulfur compounds. Hydrogen sulfide is heavier than air and therefore tends to accumulate under stagnant conditions in deep cavities, such as tunnels, vats, and cellars. It occurs in many natural gas and petroleum products and in some mining and refining operations. It is also found in sewer gas that results from bacterial fermentation. The toxicity of hydrogen sulfide is attributable to both biochemical and direct initiative actions. As a surface irritant, hydrogen sulfide primarily affects the eyes and respiratory system. In tissue liquids, the gas dissociates into hydrosulfide (HS^-) and sulfide (S^{2-}) ions, which by inactivating a number of respiratory enzymes, interfere with cellular metabolism of oxygen. As a consequence, the respiratory center in the brain may cease to function, causing apnea and sudden death. Hydrogen sulfide is deadly and, like the cyanide, exceedingly rapid in its action. However, it can also be insidious, as the warning by the sense of smell is lost early because of fatigue or paralysis of the olfactory nerve from small amounts of the gas. If death does not occur because of asphyxiation, recovery is usually complete.

Osmium Tetroxide

Osmium has only limited commercial use. Its principal forms of production are as metallic osmium and osmium tetroxide, also called osmic acid. It is used principally in histology laboratories to fix and stain tissue. The second major use is in the drug industry as a catalyst in the production of steroid hormones. The properties that distinguish osmium tetroxide as a fixative are responsible for its toxicity. It reacts with lipids, nucleic acids, and proteins. The structure and function of proteins are thereby altered. As a consequence, a variety of enzymes may be destroyed. Osmium tetroxide vapors irritate the surfaces of the skin, eyes, and respiratory tract. The subject may have smarting of the eyes, lacrimation, and see halos around lights. Corneal ulcers may occur. Among all respiratory irritants, osmium vapors appear to strike with the most dramatic intensity. Depending on the degree of exposure, all strata of the respiratory system may be involved. Coughing is the most frequent symptom. More severe exposure may cause a sense of chest constriction, coupled with difficulty in breathing. Headache behind or above the eyes is relatively common.

Oxides of Nitrogen

The term oxides of nitrogen is reserved for nitric oxide and nitrogen dioxide. Both gases are by-products of a variety of combustive processes associated with high temperature. Typically they co-exist together although their relative concentrations vary widely, depending on the nature of the combustive process. Nitrogen dioxide is the more toxic of the two gases. Oxides of nitrogen are also encountered in the chemical industry, in the manufacture and use of explosives, in the production of nitrocellulose and dyes, in nitric acid "bright dip" tanks, in decomposing nitrates, and in arc welding. Oxides of nitrogen may also be encountered on the farm in silos, often with resulting serious disabling effects in "silo-filler's disease." Nitrogen dioxide unites with water to form nitric and nitrous acids. The acids react with the tissue of the respiratory tract, producing irritating nitrites and nitrates. If the dose is overwhelming, pulmonary edema, or even death, may occur.

Ozone

Ozone, an allotropic form of oxygen, is colorless, highly reactive, and unstable. Ozone is used as an oxidizing agent in the organic chemical industry; as a disinfectant for food in cold storage rooms and for water; for bleaching textiles, waxes, flour, mineral oils and their derivatives, paper pulp, starch, and sugar; for aging liquor and wood; for processing certain perfumes, vanillin, and camphor; in treating industrial wastes; in the rapid drying of varnishes and printing inks; and in the deodorizing of feathers. Ozone is irritating to the eyes and all mucous membranes. It aggravates chronic respiratory diseases, such as asthma and bronchitis and can cause structural and chemical changes in the lungs.

Phosgene

Phosgene, or carbonyl chloride, is a highly toxic, colorless, and noncombustible gas with an odor resembling that of musty hay in low concentrations. In high concentrations, it is irritating and pungent. Along with chlorine, phosgene was used as a poisonous gas in World War I. Industrially, it is used to make acid chlorides and anhydrides and for preparing numerous other organic chemicals. It can be formed spontaneously during welding, degreasing and other operations where chlorinated hydrocarbon solvents come in contact with high energy sources. Accidental exposure to phosgene may pass unnoticed before the onset of symptoms. After exposure, there may be a feeling of tightness in the chest, cough, nausea, and vomiting. There is irritation of the eyes, nose, and throat, with respiration becoming more rapid. In addition to respiratory symptoms, there may be evidence of nervous system involvement: dizziness, headache, blurred vision, mental confusion, and muscular twitching. Death may result from respiratory or cardiac failure.

Polymer Fumes

Polymer fume fever begins several hours after exposure to the heat-degraded polymer, polytetrafluoroethylene, also known as teflon. Polytetrafluoroethylene breaks down at a temperature of 250°C to 300°C. At that point it liberates a collection of aliphatic and cyclic fluorocarbon compounds. Many are powerful irritants. The disorder is characterized by brief but acute attacks of chest tightness, choking, a dry cough, and occasional rigors.

Sulfur Dioxide

Sulfur dioxide is a colorless gas with a pungent, irritating, and suffocating odor. It is found in varying degrees in nearly every industrial community, especially in areas where coal is burned. Metallurgical plants and some refining operations release many tons of the gas into the air daily, unless measures are taken to remove the material from the stacks. Exposure may occur among sulfuric acid makers, smelters, foundry workers, blast furnace operators, bleachers, cellulose workers, coke oven workers, dyemakers, petroleum refiners, and vulcanizers. Small amounts of the gas can produce nasopharyngitis, conjunctivitis, bronchitis, and partial or temporary loss of the senses of smell and taste. High concentrations can result in laryngospasm, laryngeal edema, vomiting, and asphyxia.

Vanadium Pentoxide

Vanadium pentoxide is an industrial catalyst in oxidation reactions, used in glass and ceramic glazes, a steel additive, and used in welding electrode coatings. Of all the vanadium compounds, the pentoxide is probably the most hazardous to health. Absorption of vanadium pentoxide through the lungs results in chronic toxicity, irritation of the respiratory tract, pneumonitis, and anemia. Initial symptoms are profuse tearing and a burning sensation in the eyes, rhinitis that may be accompanied by nosebleeds, a productive cough, sore throat, and chest pain. A green-black discoloration of the tongue has been described.

Animal Proteins

Serum protein has been implicated as causative agents of hypersensitivity pneumonitis. Avian proteins may be inhaled in the dispersed dried excreta of caged pigeons, chickens, parakeets, or budgerigars. The clinical conditions have been variously described as bird breeder's lung (avian droppings), snuff lung (bovine and porcine pituitary snuff), technician's lung (rat urine), and furrier's lung (animal hair).

Hypersensitivity pneumonitis has also been described in chemical manufacturing and processing. Other industries in which the respiratory disorder may occur are variable and have little in common. As new areas of exposure and subsequent diseases are described, it has become apparent that a broad variety of inhalant organic dusts are capable of inducing hypersensitivity pneumonitis.

PNEUMOCONIOSIS

Pneumoconiosis is an occupational disease of the lungs caused by the accumulation of certain inorganic and organic dusts and the reaction of pulmonary tissue to the dusts. Originally, pneumoconiosis included only particulate matter in the solid phase. In recent years, however, the definition has been broadened to include aerosols other than dusts. The lesions of the pneumoconiosis are generally divided into two major categories. The first, simple pneumoconiosis, is a disorder in which there are discrete small lesions scattered throughout the lung in various profusion and with varying degrees of fibrosis. An example of simple pneumoconiosis is silicosis or coal worker's pneumoconiosis. The second, complicated pneumoconiosis or progressive massive fibrosis, affects a small percentage of persons with pneumoconiosis. In this case, fibrotic nodules coalesce and encompass blood vessels and airways. In the past, tuberculosis was a common accompaniment of this condition.

A type of pneumoconiosis produced by dusts that are inert is known as benign pneumoconiosis. There is no pulmonary tissue reaction to deposition of the dusts, even when there are large accumulations in the lungs. There also is no functional impairment despite somewhat dramatic x-ray findings. Barium sulfate and other barium, iron and tin dusts are inert.

Silicosis

Silicosis is a pathological condition of the lungs resulting from the inhalation of particulate matter containing free silica (SO_2). It is important to distinguish between silica in the free state as silicon dioxide and in the combined state as the various silicates. Silicosis is prevalent in many industries, and of all the pneumoconioses, it claims the largest number of victims. Silicosis is generally divided into three stages: slight, moderate, and severe. The first stage, so-called simple silicosis, supervenes in a worker who has been exposed to free silica dust for many years. The onset of symptoms is marked by dyspnea on exertion, slight at first and later increasing in severity. Cough may be present and is usually unproductive. Impairment of working capacity is slight or absent. In the second stage, dyspnea and cough become established and further physical signs appear. Some degree of impairment of working capacity begins to develop. In the third stage, dyspnea progresses to total incapacity. Right heart hypertrophy and then failure may supervene. The concentrations of dust that can be inhaled without danger vary according to the nature of the dust and also to the length of time during which it is breathed. Intermittent exposure to high concentrations of dust may be more dangerous than exposure to lower concentrations of dust over a longer time period. The harder the work, the more deeply a worker will have to breathe; in consequence, that worker will inhale more dust. Individuals also vary greatly in their capacity to respond to inhaled dusts.

Acute silicosis is a form of silicosis that can occur in workers exposed to high levels of respirable silica over a relatively short period of time. The occupations in which the disease has been described are sandblasting, rock drilling, lens grinding, and the manufacture of abrasive soaps. The history is typically of progressive dyspnea, fever, cough, and weight loss. The disease is rapidly progressive, death occurring in respiratory failure.

Nodular silicosis is a respiratory condition associated with the inhalation of silicon oxide particles ranging from 0.1 to 3 µm in size stored in excessive amounts in the lungs. The particles characteristically produce fibrous nodules that are identified as "dust granulomas." The nodules develop by a peculiar fibrotic process in which fibrous tissue is laid down in concentric rings about the central core of foci of silica particles, enveloping the particles by repeated layers of fibrous tissue until they resemble the layers of an onion. The final macroscopic, round, hard, discrete fibrous nodules are usually 2 to 4 mm in size. The nodules cause many individual alveoli to be compressed and collapse. Distortion and some breakdown of alveolar septa develops, causing minute foci of emphysema to develop about the nodules. No demonstrable impairment of the lung function results because of the great functional reserve capacity of the lung.

Silicate Pneumoconiosis

Silicates have a crystal structure containing the SiO_4^{2-} tetrahedron arranged as isolated units, as single or double chains, as sheets, and as three-dimensional networks. There are also fibrous "rock forming silicates," variously described as asbestiform, elongate, fibrous, bladed, prismatic, and columnar. At present, there is no concurrence on the effects of silicates on humans by minerals not generally regarded as asbestos. Because of the diversity of silicates, the health effects of these minerals are discussed according to their mineralogical classification.

The olivine group of igneous rocks are iron and magnesium silicates with island structures of SiO_4^{2-} units. Olivine is used principally in foundries, primarily as a special sand for mold-making in brass, aluminum, and magnesium foundries. It is used as a refractory (bricks) material, in mixes for furnace linings, and as a source of magnesium in fertilizer. No disease or pneumoconiosis from olivine mineral has been observed in humans.

The kyanite group of minerals are aluminum silicates with island structures of SiO_4^{2-} units. This group of minerals has been in great demand for making high-grade refractories, such as in spark plugs, thermocouple tubing, and refractory bricks in electric and forging furnaces and cement kilns. Mild fibrosis or pneumoconiosis may result from exposure to kyanite. Silicosis is possible if cristobalite is present.

The pyroxene group forms chains of $(Si_2O_4)^{4-}$ units bound together by covalent oxygen bonds. The commercially mined pyroxenes are spodumene and wollastonite. Spodumene, a comparatively rare mineral, is one of the commercial sources of lithium. Wollastonite is used in the ceramic industry, as a replacement for nonfibrous materials in brake linings, as an insulation for electronic equipment and thermal insulation such as mineral wool, and many other uses. There is no known disease associated with exposure.

The most common clay minerals belong to the kaolinite. The structural unit of this group is the siloxane sheet, $(Si_2O)^{2-}$, which exhibits excellent cleavage, easy gliding, and a greasy feel. The leading consumer of kaolin is the paper industry where it is used to fill and coat the surface. It is also used as a filler in both natural and synthetic rubber, as a paint extender, a filler in plastics, and in the manufacture of ceramics. Kaolinosis is a pneumoconiosis produced by kaolin. The pneumoconiosis is mainly nodular or massive fibrosis of the lung. Symptoms are dyspnea on exertion and productive cough. Pulmonary tuberculosis and emphysema are often found. A benign pneumoconiosis is more commonly seen on chest radiographs. Other silicates with sheet structures are talc, bentonite, and Fuller's earth. The effects of exposure to these silicates are discussed below.

Talc. Talc is an extremely versatile mineral that has found many uses. The principal uses of talc include: extender and filler pigment in the paint industry, coating and filling of paper, ceramic products, filler material for plastics, and roofing products. The character of pneumoconiosis associated with talc exposure depends on the composition of the talc dust inhaled. When asbestos is the dominant mineral, the disease is characteristic of asbestos-induced disorders. When talc is associated with quartz, the reaction of the lung to quartz is modified, giving rise to localized fibrocellular lesions, and is called talcosilicosis. Talcosis is pneumoconiosis caused by deposition of pure talc, i.e., asbestos-free and quartz-free. Granulomas are observed in chest radiographs.

Bentonite. Bentonite is used as a foundry sand bond, drilling mud where penetrated rocks contain only fresh water, bleaching clay, and pelletizing of talconite ore. Silicosis has been reported in bentonite workers. No disease has been associated with bentonite alone.

Fuller's Earth. Fuller's earth is a porous colloidal aluminum silicate. The term is a catch-all for clay or other fine-grained earthy material suitable for use as an absorbent and bleach. The disease is pneumoconiosis, often resembling silicosis. There is no recognized disease specific for Fuller's earth.

The framework structure group of silicates is probably the most important because nearly three-fourths of the rocky crust of the earth is composed of these minerals which are stable and strongly bonded. Quartz, tridymite, and cristobalite are the three principal crystalline polymorphs and can be transformed from one to the other under different conditions of temperature and pressure. In the mixed dust pneumoconioses, the pathology depends to a large extent upon the relative proportion of quartz present in the airborne dust. Those with a quartz content of less than about 0.1% tend to develop small nodular areas in the lungs in almost direct proportion to the total amount of dust deposited. On the other hand, dusts in which the quartz content ranges from about 2% to about 18-20% of the total dust tend to produce lesions that more nearly resemble those seen in classical silicosis.

Diatomaceous Earth

A respiratory disorder caused by inhalation of amorphous silicon dioxide particles is known as diatomaceous earth pneumoconiosis. In sufficient quantities, this amorphous diatomaceous earth may produce a mild linear reticulation without clinical symptoms. After calcining, 14 to 60% of the amorphous silicon dioxide is transformed into crystalline silicon dioxide, mostly in the form of cristobalite. When stored in the lung in sufficient quantity, it produces a diffuse fibrosis. A delicate mesh of fibrous tissue produced by the thickened alveolar wall is present.

Coal Workers' Pneumoconiosis

Coal workers' pneumoconiosis is a condition in which coal dust and similar forms of carbon are stored in the lungs forming an often nondisabling anthracosis. In addition to heavy exposure to coal dust, many coal miners, particularly those mining anthracite coal, are exposed to high concentrations of silica dust. In such exposed individuals, the accumulation in the lungs of two types of dust—silica and coal—results in the development of anthracosilicosis. Many hard-coal miners develop that condition with all the characteristics of silicosis plus the results of lung deposits of coal dust. Initially, in coal workers' pneumoconiosis, the disease is manifest characteristically by the coal macule and later by coal micronodules and nodules resulting in simple coal workers' pneumoconiosis. In some cases, large lesions of 1-3 cm in diameter, or even massive consolidated lesions, develop resulting in progressive massive fibrosis.

Black Lung Disease

"Black Lung" is a legislatively defined term which encompasses the classical medical definition of coal workers' pneumoconiosis but is defined by the Federal Mine Safety and Health Act of 1977 (P.L. 91-173 as amended by P.L. 95-164), as "a chronic dust disease of the lung arising out of employment in an underground coal mine." The definition is used to cover disability primarily from chronic airways obstruction that is associated with coal mine dust exposure. Coal miners who meet any of the qualifications criteria in the Act, and if the judgment of the examining physician and administrative law judge that they have developed their condition in association with coal mine employment, may be compensated for total disability.

Caplan's Syndrome

Caplan's syndrome is a form of pneumoconiosis that may develop in a person afflicted with or susceptible to a rheumatoid disease and who has been exposed to a dust hazard. It was first described in coal miners with rheumatoid arthritis but subsequently found in other mining operations. The symptoms result from an interaction between the inhaled mineral dust in the lung and rheumatoid factor. Its most characteristic feature is larger, more rapidly developing nodules than those seen in simple silicosis. An increased prevalence of progressive massive fibrosis is seen in these individuals.

Asbestosis

Asbestos is a group of impure magnesium silicate mineral fibers with a theoretical formula of $Mg_6Si_4O_{10} \cdot (OH)_8$. There are two major types of asbestos fibers, amphibole (short and straight) and serpentine (long, flexible, and curly). Categories defined on the basis of commercial use are chrysotile (serpentine), amosite, actinolite, anthophyllite, crocidolite and tremolite (amphiboles). Only amosite, chrysotile, and crocidolite are of economic importance. Chrysotile is basically a sheet silicate mineral rolled into itself to form a hollow tube. This tube constitutes the basic fibril of chrysotile. All amphibole asbestos types are similar in crystal structure: they consist of double chains of linked silicon oxygen tetrahedra between which metallic ions are sandwiched. More than 90% of all asbestos used in the United States is of the chrysotile variety. Asbestos cement products constitute the major use of asbestos followed closely by floor products or materials used in the construction industry. Materials containing asbestos have been extensively used in construction and ship-building for purposes of fire-proofing and for decoration.

Asbestos fibers vary considerably in size. Amosite fibers are the largest. The smallest fibers are too fine to be seen with an optical microscope and must be viewed through an electron microscope. The small fibers may be inhaled from airborne dust to form residues in lung tissue. Inhalation of asbestos fibers is associated with a number of health problems including asbestosis, lung cancer, mesothelioma, and pleural plaques. In the lung of asbestos-exposed persons may be found the presence of asbestos bodies, or ferruginous bodies, which are inhaled asbestos fibers that have become encased in a deposit of protein, calcium, and iron salts in the lung tissue. It is believed that the coating is produced by pulmonary macrophages in order to isolate the fibers and prevent them from becoming the source of a fibrosing action in the lung tissue.

Asbestosis is the name of the pneumoconiosis produced by the inhalation of asbestos fibers. It is characterized by diffuse interstitial fibrosis of the lung parenchyma, often accompanied by thickening of the visceral pleura and sometimes calcification of the pleura. The risk for asbestosis seems to vary with the amount of asbestos fiber regularly inhaled, other life-style factors such as cigarette smoking, and the type of asbestos to which the worker was exposed. One study indicates that crocidolite fibers present the greatest risk and chrysotile fibers the least risk.

Fiberglass

Glass fibers are manufactured from a melt of a batch containing silica, lime-stone, aluminum hydroxide, soda ash, and borax. Fiberglass is produced by drawing or blowing the molten glass into fine fibers that are flexible but retain the tensile strength of glass. The risk from inhalation of respirable fibers is not fully evaluated. Among the hazards to workers are those due to abrasions. Some workers may develop dermatitis from the mechanical action of the fibers on the skin, and a small percentage of them may develop an allergic dermatitis from the binder. Fiberglass is capable of producing a mechanical, transitory skin irritation characterized by a maculopapular eruption. It usually is noted at pressure points, such as around the waist, collar, and wrists. This temporary irritation usually begins to decrease within 3 to 5 days after beginning work with the material. Workers experience no lasting adverse effects once they are removed from exposure to the material.

Man-Made Mineral Fibers

Man-made mineral fibers are those made from glass, natural rock or any readily fusible slag. They differ from naturally occurring fibers, such as asbestos, which are crystalline in structure and differ chemically. Man-made mineral fibers are glassy cylinders and can never split longitudinally. They only break across. As they are destroyed, they form fragments that no longer have the character of fibers. The structure of asbestos fiber is totally different. It is always present as bundles, never as a single fiber. Individual fibrils may be as small as 2 to 30 nm in diameter. Because of the size of individual asbestos fibers, a bundle of asbestos fibers of the same diameter as mineral fibers would contain almost one million fibers. The health effects of man-made mineral fibers may be different depending on the length and diameter of the individual fibers. Some studies indicate that no significant changes are observed in chest radiographs or pulmonary function tests of workers involved in the production and use of man-made mineral fibers. However, there is some evidence that pre-existing respiratory conditions, such as asthma and bronchitis, may be aggravated by exposure to man-made mineral fibers. Fibrosis is not a hazard of man-made mineral fibers and post-mortem examinations of the lungs of persons who have worked with man-made mineral fibers show no significant difference from those of the lungs of city dwellers who have had no special exposure to man-made mineral fibers.

Natural Fibers

"Natural fibers" is a term applied to any flexible filamentous substance with a length that is many times that of the diameter. Natural animal fibers include those of wool and silk. The wool category is sometimes broadened to include alpaca, camel, goat, and mohair fibers. Silk is the only natural fiber that occurs as a continuous filament. The more important vegetable fibers are cotton, flax, hemp, jute, ramie, and sisal. Vegetable fibers are sometimes subdivided according to whether they are derived from the stem, leaf, fruit, or seed of the plant. Flax, hemp, jute and ramie are classified as bast fibers, obtained from plant stems. Sisal is a leaf fiber. Fruit or seed fibers include cotton, kapok and the coir fibers of coconut husks.

Exposure to vegetable fibers is a major occupational hazard. Of the three vegetable fibers associated with respirable dust exposure during processing, cotton is the predominant textile fiber, followed by flax which is woven into linen, and soft hemp, traditionally used for rope and net making, but now largely replaced by synthetic fibers.

Byssinosis. Byssinosis is the generic name applied to acute and chronic airways disease among those who process cotton, flax, and hemp fibers. In the textile industry, it is also known as brown lung disease. It is complicated by the presence of foreign materials and microorganisms, such as molds and fungi, that collect on fabrics. The acute response is characterized by a sensation of chest tightness upon return to exposure following a holiday or weekend break. This symptom is often accompanied by a cough, which becomes productive with time, and occasionally by shortness of breath. Measurement of lung function upon return to exposure often reveals modest decreases in expiratory flow rates over the working shift. For most affected individuals, these findings will diminish or disappear on the second day of work. With prolonged exposure, both the symptoms and functional changes become more severe. Dyspnea becomes the prominent complaint while decrements in expiratory flow rates over a work shift are often marked, and clear clinical and physiological evidence of chronic obstructive lung disease emerges.

METAL DUSTS

A group of occupational lung disorders is due to the inhalation of a variety of different inorganic materials, mainly metals of one sort or another. While the total number of exposed workers at risk may be large, the actual number of individuals employed in any single industry or exposed to any specific agent may not be large. Consequently, occupational health information is not as readily available as it is in larger industries where health and safety programs are better recorded. Another consideration is that many of the mineral dust pneumoconioses cause only chest radiographic alterations but are not associated with medical disability. Because of this, there is less interest or impetus to monitor these pneumoconioses.

An *aerosol* is airborne solid or liquid particles produced by grinding, cutting, hot operations of forging or welding, painting and spraying, extraction and crushing, shaping, and chemical reactions. Aerosols can occur in the form of dusts, fumes, smokes, or mists and fogs, according to their physical nature, their particle size, and their method of generation. *Dusts* are usually produced by mining or ore reduction, resulting in disintegration of solid materials. *Fumes* are produced by such processes as combustion, distillation, calcination, condensation, sublimation, and chemical reactions. They form true colloidal systems in air. *Smokes* are colloidal systems produced by incomplete combustion. *Mists* and *fogs* cover a wide range or particle sizes and are considered to be primarily liquid.

Aluminum

The metal of aluminum is never found in the elemental state. Its primary sources are the ores cryolite and bauxite. Most hazardous exposures to aluminum occur in smelting and refining processes. Aluminum is mostly produced by electrolysis of bauxite, Al_2O_3, dissolved in molten cryolite, Na_3AlF_6. Aluminum's effect on the lungs differs depending upon the composition of the inhaled aluminum-containing dust. Shaver described a series of cases of pulmonary disease in a group of furnace workers processing bauxite. Shaver's disease, as this occupational disease has been named, is a special form of silicosis involving ultramicroscopic silica particles, fume in character, that produce an immediate, intense irritant reaction in the bronchioles and alveoli when they are inhaled. The condition leads to acute bronchiolitis and diffuse fibrosis. Although silica is present in crude bauxite only as a 5 to 7% impurity, crystalline free silica comprises 30 to 40% of the furnace fumes.

Barium

Baritosis is a benign pneumoconiosis that results from the inhalation of dusts of barium sulfate or barium ores. The main ore of barium is baryte. Baryte is used principally as a constituent of lithopone, a white pigment employed in the manufacture of paints. It is also used as a filler in textiles, rubber, soaps, cements, and plasters. Barium is highly insoluble and radiopaque, which allows it to be used safely as a radiographic contrast medium. The inert dust of barium compounds in nonfibrogenic and baritosis is not associated with any respiratory symptoms or functional impairment. However, the radiographic appearances are quite striking. The deposits of barium appear as multiple, dense, small, rounded opacities.

Beryllium

Inhalation of any beryllium compound is potentially hazardous. It can produce either acute or chronic beryllium disease. Acute beryllium disease, the acute response to inhaling toxic beryllium compounds, is defined as disease which lasts less than one year, occurs during exposure to beryllium, and includes any of the following: nasopharyngitis, tracheitis, bronchitis, pneumonitis, dermatitis, and conjunctivitis. Ulceration may also be present, and nasal septum perforation can occur. For acute disease, the more soluble beryllium compounds, including beryllium fluoride, beryllium sulfate, and ammonium beryllium fluoride, have been implicated as the cause of both upper and lower respiratory abnormalities. In addition, acute pneumonitis has been associated with beryllium oxide, carbide, oxyfluoride, hydroxide, and zinc beryllium silicate. Chronic beryllium disease is caused by inhalation of the fumes of the metal or an alloy containing beryllium, lasts longer than one year, and usually causes both systemic and pulmonary abnormalities. Skin lesions may develop and the lesions are reddish, papulovesicular, and pruritic. Radiological pattern is non-specific, showing an image that resembles sarcoidosis, tuberculosis, mycosis, or other lung disease. Reactions to beryllium are believed to involve the immune system through formation of an antigen by a beryllium ion combining with a protein or other natural body substance. The term berylliosis should not be used as it implies two false conclusions: (1) beryl ore itself causes disease, and (2) beryllium disease is similar to pneumoconiosis.

Cadmium

Several forms of cadmium are hazardous to workers. All occur as respirable dusts, and the metal also vaporizes if heated. On an equal weight basis, cadmium vapor is considered more toxic than the dust. Most cases of acute cadmium intoxication have been associated with welding, soldering, or silver brazing. The manifestations of toxicity are chiefly respiratory. The onset of symptoms may be delayed several hours or until the worker has left the scene of exposure. The clinical picture may simulate that of an acute infection, or be mistaken for metal fume fever, particularly among welders. Slight exposure is attended by drying and irritation of the upper respiratory tract. Pulmonary edema may occur within hours of severe exposure and persist for days or weeks. Fatality following pulmonary edema is well documented. While recovery from edema generally appears to be complete within weeks, shortness of breath and impaired pulmonary function persist for years in some instances. With chronic exposure to cadmium, nasal passages become inflamed, and there is loss of the sense of smell owing to damage to the olfactory nerve. The teeth show yellow discoloration. Chronic exposure to cadmium, however, does not seem to result in any major hazard for the lung.

Hard Metal Disease

Hard metal disease is an occupational lung disease in which the respiratory system is adversely affected by exposure to tungsten carbide. Tungsten carbide is produced by blending and heating tungsten and carbon in an electric furnace, and then mixing in a ball mill with cobalt to form a matrix for tungsten carbide crystals and other metals such as nickel, chromium, and titanium. Because tungsten carbide is extremely hard and resistant to heat, it is used in the manufacture of metal-cutting tools, dental drills, and bearings. There are two major types of pulmonary reactions that occur among workers exposed to tungsten carbide: an asthma syndrome and diffuse interstitial pulmonary fibrosis. Early symptoms are cough and scanty mucoid sputum. Later, the worker complains of shortness of breath, which worsens progressively. Pulmonary function measurements reveal reduced lung volume. The pattern is that of classical restrictive disease without significant airways obstruction. Chest radiographs may show a fine reticular nodular pattern, and the heart outline is often blurred, as in asbestosis. Fine honeycombing may develop, as in other pulmonary fibrosis. The disease is seldom seen in workers with less than 10 years exposure to tungsten carbide materials.

Iron

Inhalation of metallic iron or iron compounds causes siderosis. Siderosis is a relatively benign pneumoconiosis, characterized by large accumulations of inorganic iron containing macrophages in the lungs with minimal reactive fibrosis. In its pure form, the condition probably does not progress to true nodulation as seen with silicosis and is usually asymptomatic. It is known chiefly for the abnormal changes produced on chest radiographs. Siderosis is seen in its purest form in arc welders, oxyacetylene cutters, and silver finishers. During arc welding and oxyacetylene cutting, iron is melted and boiled by the heat of the arc or torch. The iron is emitted as particles of ferrous oxide which are immediately oxidized to ferric oxide and appear as blue-gray fumes. Prolonged inhalation of these fumes can lead to siderosis. Silver finishers use what is known as jeweler's rough to polish their unfinished wares. The rouge is composed of iron oxide and is often applied with a buffer that generates a cloud of small iron and silver particles. When iron is inhaled in conjunction with other fibrogenic mineral dusts, pulmonary fibrosis results. This is referred to as mixed dust pneumoconiosis. Most of the affected miners have worked in the mines for more than 20 years. The symptoms and signs are relatively nonspecific. The miners often complain of shortness of breath, cough, and reddish-brown sputum. The shortness of breath is worst in miners who have massive fibrosis. The amount of fibrosis is in general related to the free silica content.

Manganese

A respiratory condition associated with the inhalation of particles of manganese compounds is known as manganese pneumonitis. The effects of manganese on the nervous system are well established. However, a high incidence of bronchitis and pneumonia has been reported in a group of workers involved in the manufacture of potassium permanganate. Manganese pneumonitis is slow to respond to treatment but apparently leaves no permanent damage.

Metal Fumes

Metal fume fever is a common occupational disease in environments where workers are exposed to the fumes of certain metals, as in foundries, rolling mills, welding operations, galvanizing operations, and molten metal processing. It is characterized by a feverish reaction to the inhalation of finely divided particles of metallic oxides. While zinc, copper, and magnesium are the chief offenders, cadmium, iron, manganese, nickel, selenium, tin, and antimony are responsible in some instances. The disease has an acute onset, and although there is no chronic form of metal fume fever, repeated episodes occur. The symptoms may develop in a new worker on his first day on the job and also in experienced workers on reporting to work after a weekend break, hence the popular term of "Monday morning fever." Metal fume fever symptoms include a thirst and a metallic taste sensation. There is usually a time lag of several hours between exposure and the onset of symptoms. Later, the worker has rigors, high fever, muscular aches and pains, headache, and a generalized feeling of weakness. There may be nausea, vomiting and mental disturbances, usually marked by agitation. The worker sweats profusely, and the condition is often mistaken for influenza. The diagnosis of metal fume fever is dependent upon the worker's occupational history. There is no recognized treatment of the disease.

Mercury

Three chemical forms of mercury pose occupational hazards: elemental mercury, inorganic salts, and organic salts. Liquid elemental mercury vaporizes readily at ambient temperatures. Exposure by inhalation occurs with both the vapor and the inorganic salts as dusts. Most cases of acute intoxication are accidental, e.g., following rupture of a large mercury-containing receptacle in a confined space. If the vapor has been inhaled, the clinical picture will generally reflect injury to the lung (chest pain, cough, shortness of breath) plus general toxemia (fever, chills, profound weakness, anorexia, and joint pain). In nonfatal cases, recovery is rapid and may be complete within 24 h. Chronic exposure to mercury vapor will affect the central nervous system. The clinical picture is termed "erethism". Headache and various personality changes are described, including increased irritability, depression, paranoia, insomnia, and loss of memory and mental acuity. Motor disturbances also occur. Tremors of the limbs, particularly of the hands, are often an early sign of chronic intoxication. Muscular coordination can become impaired. Other systemic effects have been discussed elsewhere.

Tin

Exposure to dust or fumes of inorganic tin may cause a benign pneumoconiosis known as stannosis. Stannosis produces distinctive progressive changes in the lungs as recorded on chest radiographs as long as exposure continues, but there is no fibrosis and no evidence of disability. Because tin is radiopaque, early diagnosis is possible.

Titanium

Titanium pneumoconiosis is a respiratory disorder caused by the inhalation of dust of a titanium compound. Titanium oxide is used as white pigment in the manufacture of paint. Titanium carbide finds extensive use in the manufacture of tools. There is some evidence that titanium oxide may produce radiographic abnormalities similar to those seen following the inhalation of iron and tin dusts. However, the condition is relatively benign, and there is no associated pulmonary impairment.

ORGANIC DUSTS

The inhalation of organic dusts may lead to two distinct pulmonary responses. First, and more common, is occupational asthma. It occurs most commonly in atopic subjects, i.e., susceptible individuals, and is characterized by changes in the airflow resistance in the conducting system of the lungs. The second and less common type of reaction is known as hypersensitivity pneumonitis. It affects the lung parenchyma, namely, the respiratory bronchioles and alveoli and does not appear to be related to atopy.

Occupational Asthma

Asthma is a disease characterized by an increased responsiveness of the airways to various stimulants and manifested by slowing of forced expiration which changes in severity either spontaneously or with treatment. Occupational asthma is a disorder with generalized obstruction of the airway, usually reversible, and caused by the inhalation of substances or materials that a worker manufactures or uses directly, or that are incidentally present at the work site. Individuals with occupational asthma complain of tightness of chest, nocturnal cough, wheeziness, and shortness of breath. Initially, these symptoms occur only while the individual is at work but later they may persist at home and on weekends. Workers who are atopic are more prone to develop occupational asthma and may do so with a relatively short exposure. Nevertheless, normal individuals may be affected until they have been sensitized.

Immunologic Mechanism

Immunologic mechanisms seem operative in occupational asthma, often affecting atopic individuals. Atopic individuals have a unique response to intranasal immunizations with certain protein and carbohydrate antigens by producing high concentrations of skin-sensitizing antibodies. This reaction is rarely observed in nonatopic (normal) subjects immunized by the same schedule. Atopic individuals develop symptoms rapidly and are forced to leave their job at an early stage because they develop disease while nonatopic individuals represent a survivor population.

Immunologic reactions are classified into four types. Type I reactions are immediate and are mediated by a specific immunoglobulin, IgE. When an atopic individual is exposed to an antigen, there is an increase in the IgE specific to the antigen. The specific IgE reacts with the cell-bound antibody to form bivalent complexes and these, in turn, trigger a series of enzymatic reactions that ultimately result in the release of mediators, such as histamine, serotonin, slow-reacting substances of anaphylaxis (SRS-A), and eosinophil chemotactic factor of anaphylaxis (ECF-A), causing the asthmatic reaction. Type II, or cytotoxic-type reaction, is mediated by reaction of antibody with surface antigen of the cell and formation of an immune complex. There is no evidence to suggest that this type of immunologic reaction takes place in occupational asthma. The Type III responses, which are related to the Arthus phenomenon and are associated with the presence of precipitins in the blood, occur several hours after the challenge. They are due to immunoglobulin, IgG. Type IV, delayed (24 to 48 h) response, or "tuberculin reaction," is cell-mediated immunity. This type of immunologic mechanism may play some role in certain types of occupationally induced asthma.

Different patterns of asthmatic reactions can be observed following bronchial provocation with specific antigens. The reactions fall into three main patterns: immediate, delayed, or combined reactions in which both immediate and delayed reactions occur. Immediate reactions occur within minutes and are relatively short in duration. The immunologic mechanism responsible for the immediate asthmatic reactions is Type I. Delayed reactions occur hours after bronchial provocation and may last for several hours or even a couple of days. There is evidence of a Type III precipitating antibody immune-complex allergic reaction. In a Type I reaction, the decline in ventilatory capacity is usually evident within 10 to 15 min. If it is a Type III response, the decline is often delayed for several hours. Skin tests in which the reaction is immediate have been beneficial in evaluating workers with occupational asthma associated with exposure to certain agents.

Many substances in the work environment can cause occupational asthma. The list of known agents continues to expand and any current list will, of necessity, be incomplete. It has been recognized that an individual with an atopic background may become sensitized to virtually any natural product of appropriate antigenicity and particle size. With respect to synthetic agents, atopy is probably of less importance although bronchial hyperactivity in atopics probably makes them more susceptible to agents causing occupational asthma, regardless of mechanism.

Substances of Vegetable Origin

Substances of vegetable origin are probably the most commonly reported causes of occupational asthma. Carpenters, joiners, and sawmill workers become sensitized to sawmill dust, fungal spores, and substances used to treat wood. Wood dusts most often implicated include cedar, oak, mahogany, keejat, African zebra wood, and western red cedar. Occupational asthma due to grain allergy is found principally in millers and bakers, although it may occur in farm workers handling grain. Outbreaks of asthma have occurred in people exposed to a prevailing wind carrying dust from neighboring mills.

Substances of Animal Origin

Substances of animal origin include animal hair, skin cells, mites, small insects, molds, dander, and bacterial and protein dust. Shepherds, farmers, jockeys, laboratory technicians, animal handlers, veterinarians, and others who are in regular contact with animals may develop occupational asthma associated with IgE antibodies. Allergic pulmonary and skin symptoms are most likely to occur in animal handlers. Occupational asthma is also associated with the use of a protein enzyme manufactured by fermentation of *Bacillus subtilis*. Workers exposed to relatively high concentrations of the dust have developed occupational asthma. Immediate skin reactions to extracts of *Bacillus subtilis* has been reported. Both immediate and late bronchoconstrictor responses were found in some individuals on bronchial challenge.

Substances of Chemical Origin

Many chemicals, both simple and complex, are associated with occupational asthma. Among inorganic chemicals, the complex salts of platinum, when given sufficiently long exposure, will result in virtually 100% sensitization. Other inorganic chemicals known to cause occupational asthma include nickel salts, chromium salts, and sodium and potassium persulfates. Organic chemicals that can cause occupational asthma include the amines, such as ethylenediamine, p-phenylenediamine, etc.; anhydrides, such as phthalic anhydride, trimellitic anhydride, etc.; pharmaceuticals, such as ampicillin, penicillin, etc.; miscellaneous chemicals such as formaldehyde, organophosphate insecticides, etc.; and of course, diisocyanates. Diisocyanates are used to manufacture polyurethane foams. Two compounds have been incriminated as causes of occupational asthma: toluene diisocyanate and diphenylmethane diisocyanate.

HYPERSENSITIVITY PNEUMONITIS

Hypersensitivity pneumonitis is a respiratory disorder that is primarily a pulmonary inflammation of the interstitial tissues. It results from sensitization and recurrent exposure to inhaled foreign substances. The disease is diffuse, predominantly mononuclear inflammation of the lung parenchyma, particularly the terminal bronchioles and alveoli. The inflammation often organizes into granulomas and may progress to fibrosis. Most individuals who develop hypersensitivity pneumonitis are exposed through their occupation.

Although a large number of organic dusts have been identified as causes of hypersensitivity pneumonitis, the pathophysiological effects are similar no matter which dust is responsible. Although not everybody who is repeatedly exposed to the antigenic organic dust develops hypersensitivity pneumonitis, a small percentage does. Similarly, while there is good evidence that a substantial proportion of the subjects who are exposed to the antigen develop antibodies, the presence of antibodies alone is not necessarily an indication that the individual has or is likely to suffer from hypersensitivity pneumonitis.

Acute Form

The clinical presentation of hypersensitivity pneumonitis depends on the immunologic response of the individual, the antigenicity and particle size of the inhaled dust, and the intensity and frequency of exposure. The most common form of clinical presentation of hypersensitivity pneumonitis is an acute, explosive episode of respiratory and systemic symptoms occurring in temporal relationship to inhalation of the offending antigen. The symptoms occur 4 to 6 h after exposure and include sudden onset fever, chills, shortness of breath, and a dry cough. Pulmonary function tests may show a decrease in forced vital capacity and a decrease in forced expiratory volume in one second, with minimal alteration of the ratio of the two variables. The radiographic appearances are those of a diffuse acinous filling process predominantly affecting the mid and lower zones. Because the episode mimics acute influenza or other viral episodes, broad-spectrum antibiotics are often used for therapy of suspected bronchitis or pneumonia. The spontaneous recovery suggests effective antibiotic therapy, but the episode recurs with reexposure. If not treated, symptoms and signs gradually regress over a period of a week to ten days.

Chronic Form

Besides the acute form of hypersensitivity pneumonitis, a chronic form exists. The chronic form occurs with repeated low-dose exposures, and although on the first few occasions there may be mild fever and chills, the continued low-dose insults are not so obviously related to occupational exposure. The afflicted individual notices the onset of dyspnea and this is accompanied by cough, malaise, weakness, and weight loss. Pulmonary function tests show a restrictive ventilatory pattern with a decrease in diffusion capacity. Chest radiographs may show diffuse interstitial fibrosis or honeycombing. Granulomas may be found dispersed in the interstitial fibrosis. Bronchiolar walls are thickened by collagen and contain lymphocyte infiltrations and their lumens are obstructed by granulation tissue. The obstruction may lead to peripheral destruction of alveoli with honeycombing.

Hypersensitivity pneumonitis may occur following the inhalation and subsequent sensitization of antigens in a wide variety of organic materials. Offending agents may be bacterial, fungal, animal proteins, chemicals, or yet undefined. Each class of these offending agents is discussed below.

Substances of Bacterial Origin

The major occupations and industries associated with hypersensitivity pneumonitis are those in which moldy vegetable compost — which by its very nature is contaminated with thermophilic actinomycetes — is handled. Thus, farmers, sugar cane workers, and mushroom compost handlers are exposed.

Bagassosis. Bagassosis is a type of hypersensitivity pneumonitis found in moldy sugar cane waste, or bagasse. Bagasse is a fibrous substance that remains after the juice containing the melted sugar is removed from the cane.

Farmer's Lung. Farmer's lung is caused by inhalation of organic dusts containing fungal spores. It is often cited as the classic example of an organic dust allergy involving the alveoli. The source of the fungal spores is found in moldy hay, grain silage or straw.

Mushroom Worker's Lung. Mushroom worker's lung is a condition similar to bagassosis and farmer's lung. It apparently is caused by inhalation of spores that thrive in the same environment as the common mushroom, which is grown on compost whose main components are straw and horse manure.

Substances of Fungal Origin

A variety of saprophytic fungi have been found to cause hypersensitivity pneumonitis. Industries in which raw wood products are handled are prone to the development of hypersensitivity pneumonitis.

Maple Bark Stripper's Disease. Maple bark stripper's disease effects lumber workers employed in stripping the bark from maple logs. The fungus grows beneath the bark of the tree. Removal of the bark during the stripping process liberates the spores into the air.

Malt Worker's Lung.   Malt worker's lung is similar to farmer's lung. During the germination of barley, malt frequently is turned on open floors, resulting in the generation of a heavy green dust containing a high concentration of fungal spores.

Woodworker's Lung.   Woodworker's lung is a progressive irreversible interstitial hypersensitivity pneumonitis that develops in workers manufacturing wood pulp. Exposure to oak dust and fine mahogany dust can produce an immediate sensitivity reaction in wood workers.

Sequoiosis.   Exposure to redwood dust can cause a hypersensitivity known as sequoiosis.

Suberosis.   Suberosis develops in workers exposed to cork dust. In chronic suberosis, fibrotic nodules and arteriolitis are present while the acute form of the disease is characterized by a granulomatous pneumonitis similar to that of farmer's lung.

Cheese Worker's Lung.   Cheese worker's lung is caused by exposure to clouds of *Penicillium casei* that are produced when mold is washed from the surface of cheeses.

REFERENCES

Beckett, W. S. and Bascom, R. (eds), *Occupational Lung Disease*, Hanley and Belfus, Philadelphia, 1992.

Gardner, D. E., Crapo, J. D., and McClellan, R. O. (eds), *Toxicology of the Lung*, 2nd edition, Raven Press, New York, 1993.

Hall, S. K. and Cissik, J. H., *Pulmonary Health Risks from Asbestos Exposure and Smoking*, in Sourcebook on Asbestos Diseases, Volume 5, edited by G. A. Peters and B. J. Peters, Butterworth Legal Publishers, Salem, NH, 1991.

Hall, S. K., Pulmonary health risk—Abnormalities found in workers handling asbestos products, *Journal of Environmental Health*, 52(3):165–167, 1989.

Merchant, J. A., Boelecke, B. A., and Taylor, G. (eds), Occupational Respiratory Diseases, DHHS (NIOSH) Publication No. 86-102, 1986.

7

Toxic Responses of the Blood

Stephen K. Hall

CONTENTS

INTRODUCTION

The circulatory system is the transport system that supplies substances absorbed from the gastrointestinal tract and oxygen to the tissues, returns carbon dioxide to the lungs and other products of metabolism to the kidneys, functions in the regulation of body temperature, and distributes hormones and other agents that regulate cell function. The cellular elements of the blood—red blood cells, white blood cells, and platelets—are suspended in plasma. The normal total circulating blood volume is about 8% of the body weight, or 5600 mL in a 70-kg man. About 55% of this volume is plasma.

In the adult, red blood cells, most white blood cells, and platelets are formed in the bone marrow which is one of the largest organs in the body. Normally, 75% of the cells in the marrow belong to the white blood cell-producing myeloid series and only 25% are maturing red cells, even though there are more than 500 times as many red cells in the circulation as there are white cells. This difference in the marrow probably reflects the fact that the average life span of white cells is short, whereas that of red cells is long.

The red blood cells, or erythrocytes, carry hemoglobin in the circulation. In mammals, they lose their nuclei before entering the circulation. In man, they survive in the circulation for an average of 120 days. The average normal red blood cell count is 5.4×10^6 cells/μL in men and 4.8×10^6 cells/μL in women. In contrast, there are normally 4,000 to 10,000 white blood cells/μL of blood. Of these, the granulocyte, or polymorphonuclear leukocytes (PMNs) are the most numerous. Most of the white blood cells contain neutrophilic granules (neutrophils), but a few contain granules that stain with acid dyes (eosinophil) and some have basophilic granules (basophil). The other two cell types found normally in peripheral blood are lymphocytes and monocytes. The platelets are small, granulated bodies and there are about 300,000/μL of circulating blood.

EFFECTS OF CHEMICALS ON BLOOD CELLS

Some chemicals can exert a toxic effect upon the hematopoietic system, resulting in abnormalities in the blood cells themselves, or on their production in the bone marrow. Classification of such chemicals is based on these two effects. Bone marrow effects are mostly hypoplasia, or underproduction of blood cells, or sometimes hyperplasia, overproduction of blood cells. In extreme cases, the bone marrow effects may be aplasia, or no production. Some of the conditions associated with hypoplasia are erythropenia (deficiency of red blood cell production), leukopenia (deficiency of white cell production), pancytopenia (deficiency of all cell production), or any combination of these.

Bone Marrow Suppression

Bone marrow suppression is characterized by a deficiency of all or some cellular elements in peripheral blood. This condition results from either a decrease in production of cells or an inability of bone marrow to manufacture adequate numbers of these cells. Bone marrow suppressors are of two groups: those that regularly produce suppression upon exposure if a sufficient dose is given (e.g., arsine, benzene, radioactive chemicals, as well as ionizing radiation); and those occasionally producing suppression that may be related to an unknown biochemical idiosyncracy of the worker (e.g., organic arsenical, gold compounds, sulfonamide, and trinitrotoluene).

Signs and symptoms of bone marrow suppression are variable and often insidious, depending upon the degree of pancytopenia. The general conditions of anemia, weakness, fatigue, irritability, and gastrointestinal distress may be present. Hemorrhage and blood loss from mucous surfaces are common and infection can become a serious problem.

Hemolytic Anemia

Hemolytic anemia is due to an increased rate of red blood cell destruction. Frequently, there is shortened life span or an increased fragility of red blood cells. Chemicals cause hemolytic anemia by at least three mechanisms: those that produce hemolysis directly in all persons if a sufficient dose is given (e.g., arsine, benzene, lead, methyl chloride, phenylhydrazine, and trinitrotoluene); those that produce hemolysis by an immune mechanism (e.g., quinine and quinidine); and those that affect people with certain genetic defects such as glucose-6-phosphate dehydrogenase deficiency (e.g., acetanilide, naphthalene, phenylhydrazine, potassium perchlorate, and sulfanilamide). Clinical tests that are available for the determination of hemolytic anemia are blood smears, cell counts, cell morphology, Heinz body accumulation, and urinary levels of erythrocyte breakdown products.

Polycythemia

Polycythemia is the presence of an abnormally high red blood cell count, an increased hemoglobin concentration, or increased hematocrit. A number of chemicals have been reported to cause polycythemia. The best-known agent is cobalt. Besides the polycythemic effect, cobalt produces reticulocytosis, i.e., an increase in the number of precursor red blood cells, bone marrow hyperplasia, and increased erythropoietic activity in the spleen and liver.

EFFECTS ON HEMOGLOBIN

Hemoglobin is the oxygen-carrying protein of the red blood cells. The globin, or protein chains, has irregularly folded conformations that enclose the heme group in a hydrophobic pocket that forms the oxygen binding site. The active site of the heme group is a Fe^{2+} ion situated in a porphyrin ring. Of the two remaining coordination bonds, one is associated with an imidazole residue from the globin chain and the remaining bond is available for reversible binding with oxygen. No ligand is known to occupy this latter site in the case of deoxyhemoglobin. The reversible binding of oxygen by hemoglobin is called oxygenation. Four oxygen molecules bind to a hemoglobin. When the hemoglobin molecule is fully saturated, all four oxygen molecules are thought to be equivalent, and any one of them may be the first to be released. The release of the first oxygen, however, will greatly facilitate the release of the second oxygen molecule. In the same manner, the release of the second oxygen facilitates the release of the third oxygen. Release of the fourth oxygen does not occur under normal physiologic conditions.

Carboxyhemoglobin

Carbon monoxide is perhaps the best-known example of a chemical agent that decreases the oxygen transport of the blood and produces hypoxia, a condition in which there is an inadequate supply of oxygen to the tissues. The mechanism whereby carbon monoxide elicits this toxic effect results from the fact that carbon monoxide is a stronger ligand for hemoglobin than is oxygen, and therefore has a stronger binding affinity. This means that carbon monoxide molecules compete more successfully than oxygen molecules for the hemoglobin binding sites that normally carry oxygen. In humans this carbon monoxide binding affinity has a value of 220 at pH 7.4 for human blood. This means that carbon monoxide binds the hemoglobin 220 times more tightly than oxygen. Thus, at equal concentrations of the two gases, the blood would contain 220 times more carboxyhemoglobin than oxyhemoglobin.

The carbon monoxide binding affinity differs among species. In the canary, for example, the carbon monoxide binding affinity is only about 110. At steady state, the canary contains less carboxyhemoglobin than a human for any given carbon monoxide concentration. In the old days, miners took caged canaries into the shafts with them as an early warning device of carbon monoxide accumulation or oxygen depletion. Canaries were useful indicators because their respiration rates and metabolic rates are much higher than those of humans. As a result, they achieve any change in equilibrium much more quickly than miners do. At high carbon monoxide concentrations, those toxic to the canary are well above those toxic for the miners, the canary dies first and warns the miner to leave. At low carbon monoxide concentrations, however, a steady state is attained in the bird first but the amount of hemoglobin deoxygenated in the canary is not as high as that deoxygenated in humans and the miner expires first warning the canary to leave. The decisive dividing line is about 0.2% of carbon monoxide in air. Below this concentration, the miner probably dies first and above this concentration, the canary dies first.

One of the most important concerns of occupational toxicology has been the relationship of atmospheric carbon monoxide level, carboxyhemoglobin level in blood, and physiologic effects. The relationship between carboxyhemoglobin percent saturated in blood and the respective toxic effects is summarized in Table 7.1. It is important to note that normal carboxyhemoglobin levels in blood are not zero, even in instances with no detectable carbon monoxide level. This small but detectable level of carboxyhemoglobin is a natural component of the blood formed from the normal breakdown of hemoglobin. It has been determined that the carboxyhemoglobin level of nonsmoking adults breathing carbon monoxide free air was between 0.3 and 0.5% in blood, and that of tobacco smokers was between 3 and 10% depending on the number of cigarettes smoked and the manner of smoking, inhaling or not inhaling. Exposure to dichloromethane can also produce increased carboxyhemoglobin levels.

The current threshold limit value (TLV) of 50 ppm carbon monoxide based on an 8-h time-weighted average exposure is designed to maintain carboxyhemoglobin level less than 10%. This is consistent with the biological exposure index (BEI) of less than 8% for carboxyhemoglobin in blood. The National Institute for Occupational Safety and Health (NIOSH), however, recommends an allowable level for carbon monoxide of 35 ppm based on an 8-h time-weighted average exposure so that carboxyhemoglobin level in blood does not exceed 5%.

TABLE 7.1   Carboxyhemoglobin Levels and Toxic Effects

% COHb in Blood	Toxic Effects
0–10	No symptoms
10–20	Tightness across forehead, possible slight headache, dilation of cutaneous blood vessels
20–40	Headache and throbbing in temples; severe headache, weakness, dizziness, dimness of vision, nausea, vomiting, collapse
40–50	Same as previous item with more possibility of collapse and syncope; increased respiration and pulse
50–60	Syncope, increased respiration and pulse, coma with intermittent convulsions and Chyne-Stokes respiration
60–70	Coma with intermittent convulsions; depressed heart action and respiration; possible death
70–80	Weak pulse and slow respiration; respiratory failure and death

The obvious and specific antagonistic to carbon monoxide poisoning is oxygen. After termination of the exposure, respirations must be supported artificially if necessary. In advanced cases of carbon monoxide poisoning, the use of oxygen hyperbaric chambers can increase significantly the rate of conversion of carboxyhemoglobin to oxyhemoglobin *in vitro*. Exchange transfusion has been used for moribund victims.

Methemoglobin

The iron in the heme of hemoglobin is normally in the ferrous state (Fe^{2+}), but it can be oxidized by certain chemicals to the ferric state (Fe^{3+}). The resulting pigment is known as methemoglobin. Methemoglobinemia can be hereditary, owing to a rare genetic defect in the hemoglobin molecule, or lack of the red blood cell enzyme that converts methemoglobin back to hemoglobin after the red blood cells take up oxygen in the alveoli. A more common form of the disorder, however, is secondary or toxic methemoglobinemia, caused by exposure to a chemical agent. Substances that cause toxic methemoglobinemia may be strong oxidizers, others may interfere with the enzymatic function of the red blood cells, and still others may block the return of methemoglobin molecules to hemoglobin. In any event, a hemoglobin molecule that has been oxidized to methemoglobin is no longer a viable oxygen transporter in the bloodstream. The result is a form of anemia. Another effect is that hemoglobin molecules still able to transport oxygen to the body tissues bind the oxygen molecules so firmly, because of the effect of methemoglobin on the oxygen dissociation curve, that tissue cells have difficulty in obtaining oxygen. Thus, the presence of methemoglobin in the bloodstream has a double adverse effect and the severity of the condition varies with the concentration of methemoglobin.

In the normal person, the natural process termed "auto-oxidation" is believed to account for the steady state small amounts of the 1% methemoglobin in blood. Generally, cyanosis becomes apparent when the methemoglobin concentration exceeds 15%, but most people do not exhibit any symptoms until about 20%. The industrial terms of "blue lip" or "huckleberry pie face" refer to the cyanotic complexion as a result of methemoglobinemia. At levels of about 20 to 70%, depending on the individual, methemoglobin weakness accompanied by dizziness, headaches, tachycardia, or dyspnea may occur.

Chemicals can convert hemoglobin to methemoglobin either directly or indirectly. Some of these chemicals are listed in Table 7.2. Direct action is associated with nitrites, chlorates, hydrogen peroxide, hydroxylamine, quinone and methylene blue. Nitrites and nitrates, which are reduced to nitrite by gastrointestinal bacteria, act by destabilizing the oxygen-hemoglobin complex, allowing oxygen itself to oxidize the iron to the Fe^{3+} state. Methylene blue accomplishes oxidation by acting as a hydrogen donor in the presence of molecular oxygen. In contrast, indirect action occurs with aniline, nitrobenzene, and other amino-, nitro-, and aryl- compounds. These chemicals are active *in vivo* in the indirect formation of methemoglobin.

TABLE 7.2   Some Methemoglobin-Generating Chemicals

Acetanilide	Naphthylamine
Aminophenol	Nitrates
Ammonium nitrate	Nitrobenzene
Amyl nitrite	Nitroglycerin
Aniline	Nitrophenol
Arsine	*p*-Aminophenol
Bismuth Subnitrate	*p*-Bromoaniline
Bromates	*p*-Nitroaniline
Chloronitrobenzene	Phenylenediamine
Dimethylamine	Phenylhydrazine
Dinitrophenol	Potassium chlorate
Dinitrotoluene	Resorcinol
Hydroquinone	Sulfonamide
Hydroxylamine	Toluidine
Methylene blue	Trinitrotoluene

Almost all methemoglobin-generating chemicals have additional toxic side effects. For example, inorganic nitrites and organic aliphatic nitrite chemicals not only cause methemoglobin formation but also vasodilation. However, it is generally agreed that a reduction in the circulating titer of abnormal pigment is a desirable therapeutic goal. If the methemoglobin is contained within intact and functional erythrocytes, the intravenous administration of methylene blue usually evokes a dramatic response. Ascorbic acid also appears to be capable of reducing methemoglobin to hemoglobin and is useful to those persons genetically deficient in glucose-6-phosphate dehydrogenase. Glucose-6-phosphate dehydrogenase deficiency is a biochemical genetic condition involving the red blood cells. This genetic deficiency is a sex-linked trait, occurring homozygously only in males and present to some degree in virtually all racial groups. The prevalence rates for this trait are shown in Table 7.3.

TABLE 7.3   Prevalence of Glucose-6-Phosphate Dehydrogenase Deficiency by Racial Groups

Racial Group	Prevalence (%)
African Americans	16.0
Filipinos	12.0–13.0
Greeks	2.0–32.0
Chinese	2.0–5.0
Jews	1.0–11.0
Scandinavians	1.0–8.0
Asian Indians	0.3
White Americans	0.1
British	0.1

Source: Office of Technology Assessment, United States Congress (1983).

Cyanide Poisoning

Cyanide inhibits cytochrome oxidase. As a result, oxidative metabolism and phosphorylation are compromised. Electron transfer from cytochrome oxidase to molecular oxygen is blocked, peripheral tissue oxygen tensions rise and the dissociation gradient for oxyhemoglobin decreases.

The treatment of acute cyanide poisoning involves, as the initial step, the generation of a safe level of methemoglobin. This can be accomplished by the administration of amyl nitrite by inhalation and the intravenous injection of sodium nitrite. Free cyanide then combines with methemoglobin to form cyanmethemoglobin which cannot transport oxygen. Although very stable, cyanmethemoglobin will eventually dissociate to yield free cyanide. Therefore the second step of treatment involves the intravenous administration of sodium thiosulfate which mediates the conversion of cyanide to the much less toxic thiocyanate that is readily excreted in urine. Methemoglobin is restored endogenously to functional blood pigment by the intracellular reductase system.

Hydrosulfide Poisoning

The hydrosulfide anion (HS^-) is as potent an inhibitor of cytochrome oxidase as the cyanide anion, and the signs and symptoms of hydrosulfide poisoning are similar to cyanide poisoning. The treatment for hydrosulfide poisoning is the same as that used for cyanide poisoning. The hydrosulfide anion also forms a complex with methemoglobin known as sulfmethemoglobin, except that no further treatment is needed to degrade the sulfmethemoglobin. Because of its ability to react with disulfide bonds under physiologic conditions, the hydrosulfide anion can be inactivated by oxidized glutathione and other simple sulfides. The sulfide so generated *in vivo* may be metabolized to sulfates.

REFERENCES

Clayton, G. D. and Clayton, F. E. (eds), *Patty's Industrial Hygiene and Toxicology*, Volume 2, 4th edition, Wiley-Interscience, New York, 1994.

Hayes, A. W. (ed), *Principles and Methods of Toxicology*, 3rd edition, Raven Press, New York, 1994.

Jandl, J. H., *Blood: Textbook of Hematology*, Little, Brown, Boston, 1987.

Klaassen, C. D. (ed), *Cassarett and Doull's Toxicology: The Basic Science of Poisons*, 5th edition, Pergamon Press, New York, 1995.

Zenz, C., *Occupational Medicine*, 2nd edition, Year Book Medical Publishers, Chicago, 1988.

8

Toxic Responses of the Liver

Randall J. Ruch

CONTENTS

INTRODUCTION

The liver is the second largest organ of the body after the skin and is the most important organ for the detoxification of foreign chemicals. It is the first organ to be exposed to potentially toxic chemicals absorbed via the stomach and intestines, and is the major organ for the detoxification of such chemicals. The liver is also important in the detoxification of chemicals absorbed into the bloodstream by other routes such as the skin and lungs.

Despite its ability to detoxify many chemical agents, toxic injury to the liver (hepatotoxicity) is one of the most common manifestations of industrial/occupational toxicity. Many solvents, degreasing agents, heavy metals, and dyes can induce liver damage in exposed workers. A variety of pharmaceutical agents and dietary foodstuffs are also hepatotoxic and can enhance the toxicity of occupational agents. Only within the last half of this century has there been recognition of the potential hepatotoxic hazards imposed by occupational agents and efforts to regulate their safe use. For example, potent hepatotoxic solvents such as carbon tetrachloride and trichloroethylene were often used in poorly ventilated rooms without adequate safety equipment or monitoring of exposure.

Hepatotoxicity is a broad term, since chemical injury to the liver occurs in many forms and by many mechanisms. Distinct anatomical and clinical forms of hepatotoxicity occur with different agents. To understand and appreciate these differences, one must understand the functions of the liver, its structure and bloodflow, and how these are related to chemical induction of liver toxicity.

LIVER FUNCTIONS

In the human, the liver resides principally on the right side of the abdomen just below the diaphragm. The liver receives blood from two sources, the *hepatic artery* and the *portal vein*. The hepatic artery contains highly oxygenated blood from the lungs and foreign chemicals absorbed into the blood through the lungs, skin, and other routes. The portal vein carries less oxygenated, nutrient-rich blood from the stomach and intestines. Blood exits the liver via the *hepatic vein*. The portal blood contains nutrients, vitamins, and foreign compounds absorbed through the stomach and intestines. Thus, the liver is the first organ to encounter potentially toxic chemicals ingested orally as well as those entering the blood via other pathways. The liver has well-developed detoxification mechanisms and is very efficient at detoxifying and excreting many chemical agents. However, when a chemical agent is incapable of being detoxified or its levels overwhelm the detoxification capacity of the liver, hepatotoxicity will result.

Specific functions of the liver include (1) maintenance of nutrient levels in the blood including glucose and cholesterol, (2) synthesis of proteins such as albumin and Factor VIII which are important in blood clotting and other blood functions, (3) formation and excretion of bile via the gall bladder, and (4) biotransformation and detoxification of endogenous and exogenous agents such as bilirubin, ammonia, steroid hormones, and foreign chemicals.

LIVER STRUCTURE

The liver can be organized into subunits based upon its anatomy or its functions. The anatomical structural subunit is the *liver lobule* which is simply a hexagon that has a *portal triad* at each corner and the *central vein* at the center (Figure 8.1). The portal triad contains branches of the portal vein and hepatic artery and a *bile duct*. This latter structure carries bile to the gall bladder as described below. The central vein drains blood into the hepatic vein for exit from the liver. Blood flows from the hepatic artery and portal vein branches in the portal triad, mixes in the *penetrating vessels*, then flows between the plates of hepatic parenchymal cells (*hepatocytes*) to the central vein. Hepatocytes perform most metabolic functions of the liver. Injury to hepatocytes can be classified anatomically in this scheme as *periportal* (around the portal triad), *midzonal*, or *pericentral* (around the central venule).

The functional structural subunit of the liver is the *liver acinus*. Rappaport and others recognized that hepatocytes perform different functions depending upon their location relative to the hepatic blood flow. They divided the acinus into three zones (Figure 8.1). Hepatocytes lying closest to the portal triad and penetrating vessels (Zone 1) are exposed to highly oxygenated, nutrient-rich blood whereas those lying closest to the central vein (Zone 3) are bathed in oxygen-poor, nutrient depleted blood. Zone 2 is an intermediate zone. In addition to these differences in blood oxygen and nutrient levels, hepatocytes also differ in metabolic capacities in the three zones. For example, cytochrome P450 enzyme levels are highest in Zone 3 and glutathione levels are highest in Zone 1. Hepatocytes in Zone 1 are also the first parenchymal cells to encounter foreign chemicals entering the liver either via the arterial or portal blood system. These differences between the three zones are involved in determining hepatocyte susceptibility to a hepatotoxic chemical.

Many other types of cells are found in the liver and participate in its functions. These cells may also be targets of hepatotoxic chemicals. The *hepatic sinusoids* which carry blood from the portal triads to the central venules, are lined by *endothelial cells*. These cells scavenge lipoproteins and denatured proteins and secrete cytokines. Also present in the sinusoids are *Kupffer cells* which are similar to macrophages. These cells ingest and degrade particulate matter, produce cytokines, and participate in liver inflammatory responses. Finally, *Ito cells* or "fat-storing cells", are located between the endothelial cells and hepatocytes. Ito cells store lipid and large amounts of vitamin A. They also synthesize and secrete collagen.

BILE FORMATION

One function of the liver that is especially important in hepatotoxicity and detoxification is the synthesis of *bile*. Bile is a yellowish mixture of many compounds including bile acids, glutathione, phospholipids, cholesterol, bilirubin, organic anions, proteins, metals, ions, and detoxified foreign chemicals. Hepatocytes produce and secrete bile into bile canaliculi which empty into larger bile ductules and bile ducts. These empty into the common bile duct which empties into the gall bladder. Bile is secreted into the small intestine after a meal or in response to other stimuli.

Components of bile are important in nutrient absorption and detoxification. Bile acids are complexed with lipids and fat-soluble vitamins before absorp-

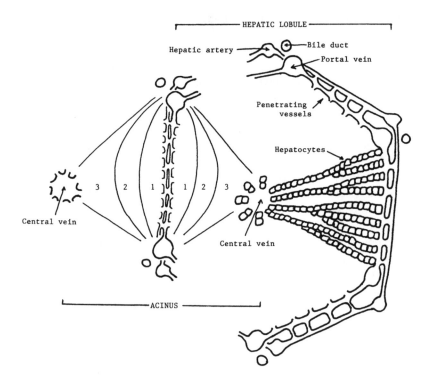

FIGURE 8.1 Structure of the liver. (Modified from Moslen, M.T., "Toxic Responses of the Liver", in *Casarett & Doull's Toxicology*, 5th edition, Klaassen, C.D., ed., McGraw-Hill, Inc., New York. With permission.)

tion. Many xenobiotics are excreted from hepatocytes in bile after their metabolism and are eventually removed from the body in the feces. Some of these metabolized agents are reabsorbed, however, and may be excreted in the urine or produce toxicity in other organs.

TYPES OF HEPATOTOXICITY

Hepatic injury depends upon the magnitude of the toxic insult, the cells that are affected, and the duration of toxin exposure (acute, chronic, repetitive, etc.). Several forms of hepatic injury are reviewed below and may occur upon exposure to occupational hazardous agents (Table 8.1).

Fatty Liver

Fatty liver (or *steatosis*) is an excess accumulation of lipid in the parenchymal tissue. Microscopically, the hepatocytes in fatty liver appear to be filled with numerous lipid droplets that push aside other cellular components. Fatty liver can occur through excess fat synthesis, decreased fat degradation, or impaired fat secretion. Fatty liver is a common and reversible acute toxic response to

TABLE 8.1 Examples of Toxic Liver Injury and Agents Responsible

Types of Injury	Examples of Toxic Agents
Fatty liver	Ethanol, carbon tetrachloride, yellow phosphorous, valproic acid, bromobenzene
Hepatocyte death	Carbon tetrachloride, bromobenzene, microcystin, ethanol, trinitrotoluene, trichloroethylene
Hepatitis	Isoniazid, nitrofurantoin
Cholestasis	Dichloroethylene, methylene dianiline, manganese, organic arsenicals, estrogens, ethanol
Cirrhosis	Ethanol, methotrexate
Blood vessel disorders	Arsenic, dacarbazine, microcystin
Tumors	Aflatoxin, arsenic, vinyl chloride, thorium dioxide

many hepatotoxicants such as carbon tetrachloride and ethanol. In fact, the consumption of a few alcoholic beverages will result in fatty liver. However, some hepatotoxic agents such as acetaminophen do not cause fatty liver.

Hepatic Death

Hepatocyte death is the end-result of excessive or prolonged exposure to hepatotoxic agents. It is often preceded by fatty liver. The mechanisms of hepatocyte death vary with the toxic agent but a common factor appears to be the reduction of cellular energy (ATP) levels due to toxic effects on the ATP-producing mitochondria in the cell. This leads to the inactivation of many ATP-dependent functions of the cell needed to maintain viability including ion and water balance. This in turn results in the accumulation of excess calcium ions and water in the cells, inactivation of cellular enzymes, and swelling and bursting of cellular components and the outer cell membrane. When this latter process occurs, the cell is dead.

Hepatocyte death can occur in a focal (limited to a few clustered cells), zonal, or panacinar or panlobular (across the entire acinus or lobule) manner. Most toxic agents cause Zone 3 hepatocyte death while fewer agents induce Zone 1 or 2 death. This may be related to the higher levels of cytochrome P450 mixed function oxidase enzymes in Zone 3 which can metabolize foreign agents to toxic components and to the higher levels of the cytoprotective molecule, glutathione, in Zone 1.

Cholestasis

Cholestasis is defined as a reduction of bile formation or impaired secretion of specific bile components. It is not necessarily associated with hepatocyte death, but may be the result of hepatocellular injury or death. Cholestasis can also occur following injury to bile duct cells or blockage of bile ducts. Several toxic agents can induce cholestasis through these mechanisms. Methylene dianiline is one agent that injures bile duct cells and causes cholestasis. Associated with cholestasis is a phenomenon known as *jaundice* in which the skin and eyes of the affected individual appear yellowish due to the accumulation of bilirubin. Normally the liver degrades hemoglobin, the oxygen-carrying pigment of red blood cells, to a yellowish compound called bilirubin which is excreted with the bile. When bile secretion or flow is reduced, bilirubin accumulates in blood and tissues such as the skin.

Liver Cirrhosis

Liver cirrhosis (or *liver scarring*) is the result of chronic toxic injury or inflammation of the liver parenchyma in which the damaged tissue is replaced with fibrous scar tissue. Repetitive exposure to hepatotoxic chemicals such as ethanol or chronic inflammation of the liver such as associated with viral hepatitis can lead to cirrhosis. The cirrhotic scar tissue is synthesized by Ito cells and other connective tissue cells in replacement of the parenchymal cells which have been killed. Normally, residual hepatocytes divide and regenerate the parenchymal tissue following a hepatotoxic injury, but with excessive injury or chronic inflammation, this process is impaired and cirrhosis occurs. The scar tissue courses through the liver and appears as tough, fibrous bands that may subdivide the parenchymal tissue into clusters of hepatocytes. This reduces blood and bile flow and greatly impairs the metabolic capacity of the liver. Cirrhosis is a common occurrence in alcoholics and is irreversible. It often results in liver failure and death of the afflicted individual.

Inflammation

Liver inflammation can occur in response to chemically-induced hepatic cell injury or as a result of viral infection (viral hepatitis). It is characterized by the activation of Kupffer cells and the presence of other inflammatory and immune system cells such as neutrophils, macrophages, lymphocytes, and plasma cells. These cells function to remove cellular debris, destroy foreign agents, produce antibodies, and repair damaged tissues. Many of these cells such as Kupffer cells produce toxic oxygen free radicals as a means to destroy foreign organisms, but which at excessive levels can also damage parenchymal cells. In addition, antibody reactions to chemical-protein adducts may occur and subsequent exposure to the chemical may elicit an immune response against the hepatic cells, leading to parenchymal cell damage.

Toxic Effects on Blood Vessels and Blood Flow

Certain industrial agents can directly damage liver endothelial cells impairing blood flow and secondarily affecting many hepatic functions. Dacarbazine, a cancer chemotherapeutic drug, is more toxic to the sinusoidal endothelial cells than the hepatocytes. Damage to endothelial cells can lead to occlusion of the sinusoid. In contrast, microcystin, a component of blue-green algae, damages hepatocytes causing them to swell and block sinusoid blood flow.

Liver Cancer

Many occupational chemicals (e.g., dyes, metals, solvents), pesticides (e,.g., DDT, aldrin, lindane), dietary factors and contaminants (e.g., alcohol, afla-toxin, nitrosamines), and pharmaceutical agents (e.g., phenobarbital, diaz-epam, clofibrate) induce liver cancer. Many of these agents do so only in experimental animals and are probably not a risk factor for human liver cancer. Agents such as aflatoxins and alcohol, however, are clearly liver carcinogens in man. The types of cancers that arise in the liver are specific to the carcino-genic agent. Most hepatic cancers develop from hepatocytes and are known as hepatocellular carcinomas. However, some carcinogens cause the formation of liver tumors from other cells. Vinyl chloride and arsenic, for example, induce tumors from endothelial cells (hemangiosarcomas). Thorium dioxide, also known as Thorotrast, was used from 1920 to 1950 as a radioactive contrast agent; this agent induces tumors from endothelial cells, hepatocytes, and biliary cells (cholangiocarcinomas).

FACTORS THAT INFLUENCE HEPATOTOXICITY

Biotransformation: Activation Pathways

The liver is the principle organ for the detoxification of ingested, inhaled, or absorbed toxic chemicals. Usually metabolism results in the generation of less toxic products that are excreted in the bile or urine. However, some chemicals enter the liver as relatively nontoxic entities but are metabolized to potent hepatotoxic products. This process of the metabolic conversion of a nontoxic agent into a toxic one is known as *biotransformation*. The vast majority of hepatotoxic compounds require biotransformation before they are toxic. Usually biotransformation is mediated by the cytochrome P450 mixed function oxidase enzymes. As discussed in other chapters, the cytochrome P450 enzyme family is large, with more than 150 members or isoenzymes. Individual isoenzymes may have broad specificity and are able to metabolize many substances whereas others have more narrow specificity and metabolize only one or a few substrates. This multienzyme cytochrome P450 system probably evolved in response to the many toxic chemicals in the prehistoric environment and diets. The toxic products of biotransformation include elec-trophilic derivatives of the parent compound which bind to nucleophilic protein (e.g., acetaminophen) or DNA (e.g., benzo(a)pyrene) and free radicals which degrade macromolecules such as lipids and proteins (e.g., carbon tetrachloride and bromobenzene).

Biotransformation: Detoxification Pathways

Enzyme systems, cofactors, and vitamins also exist in the liver to detoxify the toxic metabolites and free radicals generated by biotransformation. It is important to remember that hepatotoxicity will result when the generation of toxic moieties is greater than the capacity of the cell to detoxify them. Glutathione, a three amino acid peptide (gammaglutamyl-cysteinyl-glycine) is a very important detoxification molecule. It is present in millimolor concentrations in liver cells and has been shown to markedly reduce the toxicity of a number of compounds including bromobenzene and acetaminophen. The cysteine residue on the glutathione molecule provides an important nucleophilic thiol for the detoxification of electrophilic metabolites. The detoxification enzyme, glutathione transferase, attaches glutathione onto toxic electrophiles making the metabolite more polar and more easily excreted. Another detoxification enzyme, glutathione peroxidase, utilizes glutathione to detoxify peroxides generated by free radical reactions. Other hepatic enzymes such as glucuronyl transferase and sulfotransferase attach the polar molecules, glucuronic acid and sulfate onto electrophilic metabolites and render them less toxic and more easily excreted. Finally, cofactors such as riboflavin and vitamins C and E function as antioxidants to protect liver cells from free radical injury.

Potentiation of Hepatotoxicity

Certain chemical agents can modify the activation or detoxification of other chemicals by increasing enzyme levels, inactivating enzymes, or depleting cofactor or antioxidants. These interactions can lead to enhanced or reduced hepatotoxicity depending upon the biotransformation and detoxification systems involved. This type of interaction between chemicals may be important in the occupational setting. For example, isopropanol preexposure or ethanol consumption can enhance the toxicity of carbon tetrachloride because the alcohols increase the levels of enzymes that bioactivate carbon tetrachloride. In a similar manner, prescription medications can alter the metabolism of occupational agents and enhance or reduce toxicity.

Measurement of Hepatic Toxicity

Human hepatic toxicity may be assayed by the analysis of hepatic enzymes released into the blood from damaged liver cells. Increases in serum glutamate pyruvate transaminase (SGPT) and serum glutamate oxaloacetate transaminase (SGOT) are often seen following liver toxicity. A rise of SGPT and SGOT is also seen in obstructive jaundice, hepatitis, and metastatic carcinoma of the liver. Elevated serum alkaline phosphatase is also seen in obstructive jaundice. Reductions in serum albumin and/or globulin levels have also been used to monitor effects on hepatic protein synthesis by chemical toxins. In addition to serum enzymes, noninvasive radiographic techniques and invasive techniques such as biopsy of the liver have been used for the diagnosis of hepatocellular damage.

REFERENCES

Hayes, A. W. (ed), *Principles and Methods of Toxicology*, 3rd edition, Raven Press, New York, 1994.

Hodgson, E. and Levi, P. E. (eds), *Biochemical Toxicology*, 2nd edition, Appleton & Lange, Norwalk, CT, 1994.

Klaassen, C. D. (ed), *Cassarett and Doull's Toxicology: The Basic Science of Poisons*, 5th edition, Pergamon Press, New York, 1995.

9

Toxic Responses of the Kidney

James A. Hampton

CONTENTS

INTRODUCTION

The kidneys are complex organs which perform many functions vital to maintenance of homeostasis including: excretion of wastes; regulation of extracellular fluid volume and electrolyte composition; conversion of inactive vitamin D_3 to its active 1,25-dihydroxy vitamin D_3 form; and the systemic release of erythropoietin, renin, vasoactive prostaglandins, and kinins. A toxic challenge to the kidneys could diminish or alter any of these functions. However, normal renal function is usually assessed through the analysis of urine, the major waste product produced by the kidneys. Rapid and accurate clinical urinary tests include: volume measured as a function of time; pH; concentration of certain ions (sodium, potassium, magnesium, phosphate); glucose; aminoacids; and osmolarity. Additionally, a toxic challenge to the kidneys could result in the loss of the kidneys' ability to eliminate waste. This results in an increase in the blood urea nitrogen (BUN) and plasma creatinine concentrations. Determination of BUN is used as a general measure of renal dysfunction; however, any alteration in protein metabolism whether due to kidney or other organ dysfunction will lead to elevated BUN levels. Alterations in BUN and plasma creatinine are the classical clinical tests employed to measure decreased or impaired waste elimination. This does not indicate that waste elimination is primarily affected by a toxic challenge, only that these tests are reliable and rapid measures of decreased renal function.

More recently, elevated urinary concentrations of certain marker enzymes have been used to indicate renal dysfunction and specific sites of nephrotoxicity. For example, presence of maltase in the urine has been used as an indicator of proximal tubular dysfunction; however, many other enzymes which appear in urine following toxic insult may originate in other organs (e.g., liver) with resultant transport and elimination through the kidney. Following an acute or chronic toxic challenge, the sites of nephrotoxicity are closely associated with renal structure and function. In order to understand the mechanisms of renal toxicity and resultant nephropathy, a brief review of normal renal structure and function is warranted.

NORMAL STRUCTURES AND FUNCTIONS OF THE KIDNEYS

Close inspection of a kidney which has been sectioned sagitally reveals two major structural divisions: the granular appearing outermost *cortex* and inner striate *medulla*. Primarily located in the cortex, the basic unit of structure (functional unit) of the kidney is the *nephron* composed of the glomerulus, proximal and distal tubules, and loop of Henle. Juxtamedullary nephrons possess loops of Henle which project deep into the medullary portion of the kidney while capsular nephrons are contained primarily within the cortex. The granular appearance of the cortex is due to the sectioned proximal and distal tubules and glomeruli. The striate appearance of the medulla is due to longitudinally sectioned collecting ducts. Nephrons drain into collecting ducts which convey fluid (urine) into the renal pelvis and ureter. Groups of the collecting ducts form medullary rays, projections of medullary tissue into the cortex. The renal tissue immediately surrounding a medullary ray forms the renal lobule which is composed of nephrons and collecting ducts. In rodents, renal lobules are grouped as a single pyramidal shaped lobe which is drained by the ureter. In human, lobules are arranged into several pyramidal shaped lobes which are drained by minor calixes of the ureter. Species differences in glomerular size, length of the loop of Henle, ratio of cortex to medullary volume, and lobar architecture are seen and are closely associated with the physiologic need to produce a concentrated or dilute urine. The renal vascular supply is closely associated with the division of the kidney into cortex and medulla.

The kidney is supplied and drained by a series of arterial and venous channels which course together through and derive their names from, the surrounding renal structures. The main renal artery divides into segmental arteries in the renal pelvis which in turn branch into interlobular arteries which course between the renal lobes. At the cortico-medullary junction, interlobular arteries divide at near right angles into arcuate arteries which course along the cortico-medullary junction. Arcuate arteries give rise at near right angles to interlobular arteries which project between renal lobules toward the capsule of the kidney. Interlobular arteries give rise to intralobular arterioles which provide blood to glomeruli as afferent arterioles. Efferent arterioles drain glomeruli and branch to supply a peritubular capillary plexus which surrounds the proximal and distal convoluted tubules. The vascular elements surrounding the proximal and distal tubules serve primarily to deliver oxygen, nutrients, various substrates and waste materials to the tubule. These vascular elements also return water, reabsorbed molecules, and synthesized materials to the systemic circulation. From the peritubular plexus, capillaries merge to form an interlobular vein and venous drainage from the kidney is accomplished in the reverse order as arterial supply. The kidneys receive approximately 25% of total cardiac output. Eighty to eighty-five percent of total renal blood flow is associated with the renal cortex. Therefore, any toxic compound in the blood directed at the cortex can potentially influence cortical function. The medulla and remainder of the kidney receive approximately 10 and 5%, respectively, of total renal blood flow.

The vascular supply to the renal medulla is derived primarily from juxtamedullary glomerular efferent arterioles. Each efferent arteriolar branch divides and forms long vascular bundles termed the descending vasa recta which penetrate the medulla to varying depths. Side branches arise from the descending vasa recta to form a capillary plexus which is closely associated with and supplies oxygen and nutrients to the medullary elements of the nephron (descending and ascending tubular limbs, loop of Henle) and medullary portion of the collecting duct. Long vascular bundles termed the ascending vasa recta drain the medulla and join interlobular or arcuate veins. By entering and draining the medulla through the cortico-medullary junction, the capillary plexus performs the additional vital task of maintaining the osmotic gradient essential for the functioning of the counter current exchange and multiplier systems needed to produce concentrated urine. The processes involved in the production of concentrated urine are complex and require brief summary in order to understand the mechanisms of renal toxicology.

The afferent arteriole delivers blood under high pressure to the glomerulus where selective filtration occurs. Hydrostatic pressure is the primary force responsible for the production of the glomerular filtrate. The filtration barrier consists of (1) the fenestrated endothelial cells lining the glomerular capillary, (2) epithelial podocyte foot processes, and (3) the intervening basement membrane derived from both the glomerular endothelium and supporting epithelial podocytic cells. Since this barrier is composed of cells with holes (fenestrated endothelium) or slits (podocyte foot processes), the selectivity of glomerular filtration is accomplished by the fused basement membranes. Ultrastructurally, the basement membrane has three components. The central layer is electron-dense, contains fibers (type IV collagen), and is termed the lamina densa. The two outer most layers are more electron-lucent and are termed lamina rara interna, and externa. Materials which pass through the filter are limited to small molecular size and net molecular charge.

The glomerular filtrate is delivered to the proximal convoluted tubule where nearly 90% of water, electrolytes, glucose, and amino acids are reabsorbed. Selective elimination (secretion) of waste products and certain organic compounds also occurs by the cells composing the straight portion of the proximal tubule (pars recta). As the filtrate passes through the tubule it remains isotonic with the blood. Even with 90% reabsorption accomplished within the proximal tubule, the filtrate still requires further concentration to prevent the elimination of a large quantity of dilute urine. This is accomplished by the proximal descending and distal ascending tubular limbs, the loop of Henle, and the collecting duct. The major purpose of the descending and ascending limbs of the proximal and distal tubules and loop of Henle is to create a hypertonic gradient within the surrounding tissue fluid which increases in osmotic strength as it nears the medullary papilla. In the ascending limb, sodium is actively transported out of the tubule (with passive diffusion of chlorine) and retention of water. As a result of this process, tissue fluid surrounding the tubule becomes increasingly hypertonic, while as the filtrate ascends the tubule, it becomes hypotonic to the surrounding environment. The cells which compose the walls of the descending limb allow the passage of water, which results in the filtrate remaining isosmotic with the surrounding tissue fluid. When the filtrate is delivered to the distal convoluted tubule, it is hypotonic to the surrounding tissue fluid. As the filtrate moves through the distal tubule, water and sodium are removed and hydrogen, potassium, and ammonia enter the tubule with the net production of an isotonic filtrate that reaches the collecting duct. When the filtrate enters the collecting duct, further concentration of urine results. As the collecting duct enters the medulla and passes through the osmotic gradient established by the loop of Henle and ascending and descending tubular limbs, water is removed and the filtrate is concentrated. Antidiuretic hormone (ADH), produced by the posterior lobe of the pituitary, exerts its effect on the cells composing the collecting duct. Decreased levels of antidiuretic hormone result in less water being removed from the distal tubule and collecting duct with the net production of dilute urine. Because of the anatomical structure and functional activity of the kidney briefly described above, nephrotoxins may act on a specific area within the kidney.

EFFECTS OF NEPHROTOXINS

Since cortical tissue receives up to 80% of renal blood flow, a nephrotoxin in the blood will be directed primarily to the cortex. Although a much smaller percentage of the blood in the kidney is directed to the medulla, due to the functioning of the counter current exchange and multiplier systems a nephrotoxin could be trapped and concentrated in the medulla and affect medullary tissue. In general, a nephrotoxin could act on the vascular elements supplying a nephron, or any portion of the nephron directly. The effects of a nephrotoxin range from a minor and transitory alteration in cell transport activity (glucoseurea, aminoacidurea); secretory failure; failure to concentrate urine (polyurea); to cell death, necrosis and renal failure. A nephrotoxin could: (1) result in vasoconstriction and ischemia, with decreased glomerular filtration and tubular transport; (2) affect the glomerulus directly, with resultant filtration dysfunction (increase/decrease); or (3) affect tubular (proximal/distal/loop of Henle/collecting duct) activities, with resultant resorption/barrier/secretory dysfunctions.

Most nephrotoxins appear to exert their effects on the pars recta of the proximal tubule or the convoluted portion of the proximal tubule. The loop of Henle and collecting ducts appear to be affected by chronic administration of aspirin and phenacetin (analgesics). Other nephrotoxic materials such as fluoride ion appear to act on medullary tissue. The distal convoluted tubule occupies a relatively small percentage of a nephron and appears not to be selectively damaged by most nephrotoxins. However, compounds such as amphotericin (an antibiotic) have been shown to affect urine acidity, a distal convoluted tubule function. As previously discussed, the glomerulus selectively filters the blood and limits passage of substances including toxicants on the basis of molecular size and net charge. As substances (water, glucose, aminoacids) are selectively reabsorbed, nontoxic concentrations of compounds in the blood can reach toxic levels intracellularly or within the tubule and influence renal function. Similarly, toxic substances which are actively secreted by proximal tubular cells could reach high concentrations intracellularly or within the tubule resulting in toxicity. The ability of the kidneys to compensate for loss of cellular mass is well established. This requires careful distinction between chronic and acute renal injury.

Acute toxic insults to the kidney may alter renal function extensively; however, if the injury is sublethal and of short duration, recovery to normal function can occur rapidly. Similarly, chronic low dose administration of nephrotoxins over a long period may result in extensive renal damage; however, marked changes in renal function may not be detected due to the compensatory capacity of the kidney. Measures of renal function made some time after acute toxic insult may not indicate abnormalities. Following chronic toxicity, no change in renal function may be detected until the kidney has exhausted its compensatory capacity. In general, toxic substances may indirectly influence renal function due to general systemic effects including hypotension or hypertension as well as hormonal and neuronal imbalance. The remainder of this chapter will only focus on toxicants which directly influence renal function. Many nephrotoxic compounds exert the effects directly or require metabolic activation within the kidney. Other potential toxic agents are activated in other organs (e.g., the liver) and transported to the kidney where they exert their effects directly or are further metabolized by the kidney and become nephrotoxic.

HEAVY METALS

Exposure to most heavy metals results in renal toxicity. Histopathological examination of kidneys following sublethal heavy metal exposure reveals necrotic proximal tubules containing proteinaceous casts. Clinical findings correlate well with the failure of the kidney to reabsorb glomerular filtrate (polyurea, glucosurea, aminoacidurea). Increasing doses may cause renal failure, renal necrosis, elevated BUN and death. In addition, ischemia produced by vasoconstriction with resultant decrease in glomerular blood flow and filtration may be involved in cellular toxicity induced by heavy metals. Following low dose exposure, heavy metals can be detected in renal tissues prior to signs of toxicity. Renal tissues may be spared by a protective mechanism involving cellular lysosomes that occurs in low-dose heavy-metal toxicity. Cellular damage may result when these protective mechanisms have been exhausted by further heavy metal exposure.

Mercury

A potent renal toxin, mercury is introduced into the body in three forms: elemental mercury, inorganic mercury, and organic mercury. All forms of mercury are nephrotoxic. At low doses mercuric chloride is toxic primarily to the pars recta of the proximal tubule, with little effect on the more convoluted segment. Glucose reabsorption is not affected; however, the ability of proximal cells to secrete organic ions is diminished. At higher doses, all of the proximal tubule is affected. The mechanism involved in mercury toxicity primarily involves combination with sulfhydryl groups and inhibition of oxidative enzyme systems.

Platinum

Platinum nephrotoxicity is frequently a complication of cisplatin (an antitumor drug) therapy. Similar to nephrotoxicity induced by mercury (damage to pars recta of the proximal tubule), platinum compounds alter glomerular filtration as well as affect distal tubule and collecting duct functions. Pathophysiological findings are associated with proximal tubular dysfunction, ion loss (magnesium), as well as production of copious dilute urine. The inability of the kidney to produce concentrated urine may be due to cisplatin inhibition of ADH release by the pituitary. Onset of renal lesions following cisplatin therapy is slow and renal lesions may persist following cessation of treatment.

Cadmium

Cadmium toxicity results in enhanced synthesis within the liver of the metal binding protein, metallothionein. Metallothionein-cadmium complex spares cadmium toxicity to certain organs (e.g., testis), but enhances cadmium nephrotoxicity. This is accomplished by the increased renal uptake of the metallothionein-cadmium complex over the free cadmium ion. Nephrotoxicity of metallothionein-cadmium complex is localized to the proximal tubule. Pathophysiological findings are associated with reabsorption defects (glucoseurea, aminoacidurea), presence of low and high molecular weight proteins in the urine, and ion loss (phosphate). Presence of high molecular weight proteins in the urine suggests glomerular filtration defects with cadmium toxicity.

Other Metals

Renal toxicity has also been observed with chromium (primarily as potassium dichromate), arsenic, gold, iron, antimony, thallium and lead. At low doses, chromium produces generalized ischemic and tissue damage visible on the surface of the kidney as well as proximal convoluted tubular necrosis. Similar to the toxic effect of mercury, at higher doses, the entire proximal tubule is affected.

HALOGENATED HYDROCARBONS

Generally, halogenated hydrocarbons are both hepatotoxic and nephrotoxic, although considerable variability of the nephrotoxicity is seen due to species, sex, and strain differences. Production of nephrotoxic halogenated hydrocarbon metabolites occurs through a number of mechanisms. They may be produced directly by the kidney; or produced in the liver and transported to the kidney with resultant damage. A non-nephrotoxic metabolite may be produced in the liver then transported to the kidney, where further biotransformation results in the production of the ultimate nephrotoxin.

Carbon Tetrachloride and Chloroform

Both of these halogenated hydrocarbons produce nephropathy primarily associated with proximal tubular damage as well as secondary defects to the entire nephron. At low doses, pathophysiological findings are associated with proximal tubular necrosis, glucosurea, aminoacidurea, proteinurea, and decreased secretion of organic ions. Glomerular or distal tubular functions are not apparently affected. At higher doses, renal necrosis, renal failure, elevated BUN, and death result. In the liver and kidney, evidence exists that halogenated hydrocarbons are metabolized by mixed function oxidase enzyme systems which generate reactive metabolites that induce tissue damage. Experimental evidence indicates that following administration of radiolabelled carbon tetrachloride or chloroform, most radioactivity is associated with proximal tubular cells and centrolobular hepatocytes. Correlating with the major sites of tissue damage, proximal tubular cells of the kidney and centrolobular hepatocytes of the liver possess the highest organ concentrations of mixed function oxidase system enzymes. Considerable strain and sex differences are seen in chloroform nephrotoxicity. For example, in certain strains of male mice, low doses of chloroform produce renal necrosis with no apparent liver toxicity.

Hexachlorobutadiene

This halogenated hydrocarbon is a widespread environmental pollutant which is a potent nephrotoxin with little hepatic toxicity. Considerable species and strain differences are seen in hexachlorobutadiene (HCBD) nephrotoxicity. In rats, at low doses HCBD nephrotoxicity is limited to the pars recta of the proximal tubule, while in mice the entire proximal tubule is affected. At low doses, pathophysiological findings are associated with failure to reabsorb tubular filtrate (glucosurea, proteinurea, aminoacidurea), decreased secretion of organic ions, and necrotic proximal tubules. At high doses, renal failure, renal necrosis, increased BUN, and death result. Evidence exists that hepatic metabolism of HCBD results in the production of HCBD-glutathione conjugates. Production of HCBD-glutathione conjugates is the proposed mechanism for the sparing of hepatic tissues from the toxic effects of HCBD. However, HCBD-glutathione conjugates are transported to the kidney where further renal metabolism generates the ultimate nephrotoxin with resultant tissue damage. Additionally, nephrotoxicity in the proximal tubule may be enhanced due to the ability of the tubule to concentrate the reactive metabolite.

Bromobenzene

This halogenated hydrocarbon is both hepatotoxic and nephrotoxic. In vitro evidence indicates that renal toxicity does not result when bromobenzene is added to renal tissue slices. Further evidence indicates that the hepatotoxic and nephrotoxic metabolites are generated by the liver. The nephrotoxic metabolites are then transported to the kidney where direct tissue damage results without further renal transformation.

2-Bromoethylamine (BEA)

In chronic exposure to rats, this halogenated hydrocarbon results in almost complete renal papillary necrosis. Within 24 to 48 h of exposure, pathophysiological findings include necrosis of loop of Henle and collecting duct with sclerosis of juxtamedullary glomeruli and urinary concentration defects. Capsular glomeruli are not affected since they possess short loops of Henle which may be contained within the cortex or penetrate the medulla only a short distance. Clinical findings include production of dilute urine (polyurea), ion loss (sodium, potassium, chlorine), and contraction of the extracellular volume. BEA is used extensively in animal studies to mimic the effects of analgesic nephrotoxicity, and to examine therapeutic agents which may prevent analgesic (aspirin, phenacetin, acetaminophen) medullary nephrotoxicity.

Methoxyflurane

This halogenated hydrocarbon is used primarily as an anesthetic and has been shown to produce medullary necrosis and renal failure. In both animals and humans, methoxyflurane nephrotoxicity results in a defect in the ability of the kidney to produce concentrated urine characterized by increased urinary output and decreased urinary osmolarity, ion loss, contracted extracellular fluid volume, increased BUN, and renal failure. Histologically, necrosis of the loop of Henle, ascending and descending tubular limbs and collecting duct are seen. The metabolism of methoxyflurane results in the production of inorganic fluoride and oxalate with resultant nephrotoxicity. Evidence indicated that fluoride ions induce a resistance of collecting duct cells to the effects of ADH, thus leading to the production of dilute urine.

ANALGESICS

In humans, chronic ingestion of high doses of analgesics (aspirin, phenacetin, acetaminophen) results in medullary interstitial inflammation, fibrosis, renal papillary necrosis and renal failure. Nephrotoxicity of analgesics is similar to that described for BEA. Considerable variability exists in the mechanisms of analgesic renal toxicity. Nephrotoxicity of aspirin and phenacetin may be due to vasoconstriction of the vasa recta and/or inhibition of prostaglandin synthesis and removal of vasodilator prostaglandins with resultant vasoconstriction and ischemic medullary damage. Acetaminophen and its glutathione conjugates become concentrated and trapped in the medulla by the countercurrent mechanism. Within the medulla, reactive metabolites of acetaminophen are formed with resultant medullary tissue damage. Acute acetaminophen ingestion can also cause cortical nephrotoxicity in patients with hepatic insufficiency. Normally, acetaminophen is metabolized within the liver to its nontoxic glutathione-acetaminophen conjugate. When glutathione is depleted or the pathway impaired, acetaminophen is metabolized by the mixed function oxidase enzyme system into nephrotoxic metabolites which are then transported to the kidney with resultant cortical tissue damage.

ANTIBIOTICS

Certain antibiotics (neomycin, gentamicin, tobramycin, netilmicin, kanamycin, amikacin and streptomycin) which possess a molecular composition that includes aminoglycoside side chains are known to be nephrotoxic. Nephrotoxicity of aminoglycoside antibiotics occurs primarily at the proximal convoluted tubule and glomerulus. In the glomerulus, nephropathy is associated with changes in the endothelial pore size and filtration dysfunction. In the proximal tubule, aminoglycoside antibiotics are absorbed and concentrated by tubular cells with resultant alterations in normal function. The Cephalosporin class of antibiotics produces acute proximal tubular injury by its ability to accumulate in tubular cells. In particular, cephaloridine may exert its proximal tubular toxicity through an active metabolite produced by the mixed function oxidase system. In mice and rats, inhibitors of renal mixed function oxidase enzymes inhibited cephaloridine nephrotoxicity. Cephaloridine treatments deplete glutathione levels within the cortex but have no affect on glutathione levels in the medulla, which is consistent with sites of cephaloridine nephrotoxicity. In addition, pretreatment of mice and rats with glutathione depleters enhances cortical nephrotoxicity of cephaloridine.

Tetracyclines are known to be renal medullary toxins which result in the failure of the kidneys to produce concentrated urine. Outdated tetracyclines may produce a proximal tubular nephropathy associated with glucosurea, aminoacidurea, and proteinurea. Penicillins and sulfonamides may induce an immunologically mediated renal inflammatory interstitial nephritis. Amphotericin B is known to produce lesions in both proximal and distal tubules. The effect of amphotericin on the distal tubule results in the nephropathy, which is characterized by increased permeability to proteins such as albumin. The mechanism for the increased protein filtration may result from alterations in the net charge of the glomerular basement membrane with resultant filtration dysfunction.

ENVIRONMENT CONTAMINANTS

Certain herbacides, polybrominated and polychlorinated biphenyls are ubiquitous environmental contaminants which have reached sufficient levels to pose a threat to human health.

Herbicides

Paraquat and 2,4,5-trichlorophenoxyacetic acid (2,4,5-T) are widely used herbacides that have been shown to affect renal capacity to secrete organic ions. Although not directly nephrotoxic, at sufficient concentrations, both compounds have been shown to inhibit organic ion secretion by the proximal tubule, which may influence the renal toxicity of other compounds.

Polychlorinated Biphenyls (PCB) and Polybrominated Biphenyls (PBB)

Both PCB and PBB are widely dispersed environmental contaminants which are not overtly toxic to the kidney. However, similar to the function observed in the liver, these compounds have been shown to enhance drug metabolizing enzyme systems in the kidney and thus pose a potential hazard.

Tetrachlorodibenzo-p-dioxin (TCDD)

TCDD is known to be extremely toxic to animals and humans, but has not been shown to have a direct effect on the kidney. Like PCB and PBB, TCDD enhances hepatic and renal drug metabolizing enzymes.

Mycotoxins

Produced by various fungi, mycotoxins (rubratoxin B, aflatoxin B1, citrinin, ochratoxin A) are present in contaminated animal and human foods (cereals, grains, feeds). Nephropathy produced by mycotoxins is usually associated with proximal tubular dysfunction and reduction of organic ion transport.

REFERENCES

Ham, A. W. and Cormack, D. H., *Histology*, J. B. Lippincott, Philadelphia, 1979.

Hook, J. B. (ed), *Toxicology of the Kidney*, Raven Press, New York, 1981.

Klaassen, C. D. (ed), *Cassarett and Doull's Toxicology: The Basic Science of Poisons*, 5th edition, Pergamon Press, New York, 1995.

Walsh, P. C., Retick, A. B., Stamey, T. A., and Vaughan, E. D., *Campbell's Urology*, W. B. Saunders, Philadelphia, 1992.

Williams, P. L. and Warwick, R., *Gray's Anatomy*, W. B. Saunders, Philadelphia, 1980.

10

Toxic Responses of the Skin

Randall J. Ruch

CONTENTS

INTRODUCTION

The skin is the largest organ of the body and acts as an interface between the external and internal environments. The skin functions as a barrier to external chemical agents and in maintaining the body's internal composition. With its dead outer layer, the skin is an excellent permeability barrier to certain chemical agents, but readily allows the entry of others. For many chemicals, the skin is the major route of entry. The rate of absorption through the skin is an important determinant in the toxicity of many hazardous chemicals.

The skin is the most common site of work-related toxicity. Skin diseases and chemical burns may account for nearly one-half of all occupational disease. Thus, chemical exposure of the skin warrants attention in industrial settings.

THE STRUCTURE OF THE SKIN AND ITS ROLE IN PERCUTANEOUS ABSORPTION

The skin is a complex organ, and differs in thickness, cornification, pigmentation, occurrence of hair and glands, and blood supply over various regions of the body. The skin consists of two main layers: the outer *epidermis* which contains epithelial cells arranged in layers and the *dermis* which consists of fibrous connective tissue and the blood supply. The *subcutaneous layer*, which contains fat and connective tissues, lies below the dermis.

The epidermis contains several cell types. Keratinocytes are the most numerous. Keratinocytes originate from dividing "stem" cells that lie along the bottom layer of the epidermis. Old keratinocytes are gradually pushed upward by new keratinocytes. As the keratinocytes progress towards the skin surface, they produce keratin, a tough, fibrous protein. This process is called cornification or keratinization. When the keratinocytes reach the outermost layer of the epidermis, they become flattened and die resulting in stacked layers of thin, dead cells, each filled with bundles of keratin filaments. This outer layer of dead cells is the *stratum corneum* and comprises the major permeability barrier of the skin.

Besides keratin, the stratum corneum also contains considerable amounts of lipid (fat) originating from keratinocyte membranes. Thus, the stratum corneum is highly impermeable to water and water-soluble substances, electrolytes, and certain chemicals, but more permeable to fat soluble substances.

Besides the keratinocytes and their stem cells, the epidermis also contains Langerhans cells which function in immunity, melanocytes which produce the skin pigment melanin, and Merkel cells which are specialized sensory cells. Inflammatory cells and lymphocytes may also be present in the epidermis at certain times. The epidermis lacks blood vessels.

The epidermis is separated from the dermis by the basal lamina. This boundary zone consists of a thin layer of extracellular matrix components including proteins, proteoglycans, and glycosaminoglycans. The basal lamina affords little resistance to chemical penetration.

The dermis is a loose fibrous tissue containing collagen and other fibrous components. The most numerous cell type is the fibroblast, although macrophages, mast cells, and lymphocytes are also present. The dermis is richly supplied with blood vessels and functions in the regulation of body temperature. A rich network of lymphatic ducts is also found in the dermis to drain away excess fluids.

The subcutaneous layer lies below the dermis and contains adipose (fatty) and connective tissues. Adipose tissue can serve as a storage site for lipophilic chemicals that have entered the body through the skin or other routes.

The skin also contains hair follicles, sweat glands, and sebaceous glands. The presence or absence of these appendages varies greatly dependent upon the region of the skin. Collectively, appendages make up approximately 0.1% of the cross-sectional area of the skin and these structures are probably not an important route of entry for toxic agents.

The passage of external chemicals into and through the skin is called *percutaneous absorption* and this process is an important determinant in both the local and systemic toxicity of a chemical agent. The major barrier to the absorption of a chemical into the skin is the stratum corneum. Chemicals cross this barrier by passive diffusion. Hydrophilic (water soluble) substances are normally readily excluded but may enter more easily through moist skin. Lipophilic substances enter skin more easily. For example, isopropanol penetrates the skin poorly while organophosphate insecticides and polychlorinated biphenyls enter the skin easily.

Once through the stratum corneum, a substance must pass through the living cells of the epidermis and dermis to reach the circulation. These layers offer little barrier to chemical penetration.

Since the penetration of a chemical into the skin occurs by simple diffusion, the following factors will determine the percutaneous absorption of a chemical. These are the concentration and volume of the chemical, the duration of exposure, the area of exposure, the skin temperature, the binding of the chemical to the skin, the vehicle or solvent for the chemical, and the chemical nature of the toxic substance (polar versus non-polar). In addition, the regional nature of the skin itself will affect chemical absorption. Areas with thicker stratum corneum such as the palms or soles of the feet will act as better barriers to external agents than areas with thinner stratum corneum such as the forehead or inner wrist. The water content of the stratum corneum will also affect the entry of compounds. Hydrophilic agents will pass through moist skin more readily than through dry skin.

BIOTRANSFORMATION AND EXCRETION

The skin, particularly the living cells of the epidermis, is an actively metabolizing organ capable of biotransforming many chemical agents. The principles of biotransformation discussed in other chapters apply to the skin. Skin is capable of activating non-toxic chemicals to more toxic forms, principally through the presence of mixed function oxidase enzyme systems (cytochrome P450 enzymes). The skin is also capable of detoxifying toxic agents through mixed function oxidase activity and through conjugation reactions. Biotransformation enzymes in the skin such as aryl hydrocarbon hydroxylase (AHH), which can act upon benzo(a)pyrene, are inducible as in other tissues. The skin has relatively less biotransformation activity compared to other organs such as the liver or kidneys. The skin has about 2% of the activity of AHH compared to the liver. However, the contribution of the skin to the metabolism of a chemical will depend on the nature of the chemical, the route of exposure, the distribution of the agent, and the dose. Thus, in some cases, biotransformation by the skin may play an important role in the metabolic fate and toxicity of a chemical.

In addition to its biotransformation role, the skin may serve as a route for the excretion of toxic agents. Xenobiotics may be lost from the body through shedding of skin cells (desquamation), loss of hair, or secretion of sweat and sebum. For example, metals such as arsenic, copper, and lead are incorporated into the hair, and their levels in hair may be reflective of exposure. Metals may also be excreted in sweat. In general, however, compared to urinary, fecal, and respiratory routes of excretion of toxic chemicals, the skin has a minor role.

TOXIC SKIN REACTIONS

The toxic responses that occur in the skin are dependent not only upon the chemical nature of the agent, but also upon the site of absorption and method of application. The following toxic responses will be discussed: irritant reactions, allergic contact responses, photosensitization, chemical acne (including chloracne), and other responses.

Non-Immune Cutaneous Irritation

A cutaneous irritant is an agent that produces an inflammatory response in the skin (dermatitis) at the site of contact by direct action without the involvement of an immunologic mechanism. In man, cutaneous irritation results in erythema (reddening), edema (swelling), vesiculation (blistering), scaling, and thickening of the epidermis. Histologically, the key feature of cutaneous irritation is intracellular edema (spongiosis) of the epidermal cells. Irritation of the skin is important and may account for 80% of clinically recognized human contact dermatitis.

Two forms of cutaneous irritation are recognized: *acute irritation* and *cumulative irritation*. The nature of the irritation response is dependent on the chemical and mode of exposure. Acute irritation is a local, reversible inflammatory response of the skin to direct injury by a *single application* of a toxic chemical without the involvement of the immune system. Acute irritation is produced by a great number of chemicals, most of which are highly reactive. These include strong acids, bases, and solvents. Cumulative irritation is an irritation occurring in respone to *repeated or continuous exposure* of the skin to chemicals that in themselves do not cause acute irritation. Substances producing this type of reaction are termed marginal irritants. There can be great variation between persons in degree of irritant response, both acute and cumulative. Consequently, it is difficult to predict one person's response to an irritant based upon the reaction of another individual.

A large number of tests have been developed for predicting the irritant potential of a compound. The most widely used is the *Draize test* introduced in 1944. In this test, substances are applied to the skin of the backs of albino rabbits. The substance is first dissolved in a nonirritating solvent then applied to a 1" × 1" area of the skin and covered with surgical gauze. The animals are then immobilized with the gauze taped in place and the trunks of the animals wrapped with an impervious material such as rubberized cloth to prevent evaporization of volatile substances. After a 24-h exposure period, the patch is removed and the resulting skin reactions are evaluated immediately and again 72 h later.

The Draize test has undergone modification over the years and there is dispute over its reliability. However, the test is a good predictor of strong irritants. The Draize test is often a legal requirement in the evaluation of potentially toxic chemicals. Because of public opposition to this type of testing, however, many new tests using cultured epidermal cells are being developed and validated to replace the Draize test.

Immune-Mediated Cutaneous Irritation

Immune-mediated (allergic) contact dermatitis is the result of a cell-mediated or type IV immune response. This form of contact dermatitis is important because of the specificity of the response and the low amounts of allergic material that may be necessary to elicit a response.

The development of contact allergy in the skin is characterized by three stages. After the initial exposure to a potentially allergic compound, there is a *refractory period* of a few days or longer in which sensitization does not take place. After this period, the *induction period* occurs wherein sensitivity to the antigen develops. This may require 10 to 21 days. After the individual has become sensitive to the antigen, reexposure will result in an *allergic reaction* following a characteristic delay of 12 to 48 h. Once an allergic reaction has occurred, the individual may remain sensitive to the antigen for the remainder of his or her lifetime.

Cutaneous antigenic chemicals are generally of low molecular weight. There is a great range in the antigenic potency of toxic chemicals. Strong allergens are usually lipid-soluble, protein reactive, aromatic compounds with molecular weights less than 500. Some important occupational contact allergens include metals (nickel, chromium, cobalt, and organomercurials), rubber additives (thiuram sulfides, *p*-phenylenediamine, and resorcinol monobenzoate), epoxy oligomer, methyl methacrylate and other acrylic monomers, phenolic compounds, aliphatic amines, and formaldehyde. There is a great diversity between individuals in the ability of these and other substances to elicit allergic responses. The phenomenon of cross-sensitization may also occur for two or more chemicals. This is the process whereby an individual who is sensitized to one chemical is also sensitive to a second because of its similar chemical structure.

An important means by which chemicals are identified as allergens in an individual is by *diagnostic patch testing*. In these tests, a nonirritating and nonsensitizing quantity of the suspected allergen is applied to the skin in a vehicle such as petroleum jelly and covered with a patch for 48 h. After exposure, the skin is observed for several hours for evidence of an allergic response.

A number of tests have also been developed to identify the allergic potency of chemicals. These tests all require the use of human volunteers or live animals with competent immune systems. No satisfactory cell culture tests have been developed. In these tests, the subjects are exposed repeatedly to an agent by skin painting or by intradermal injections. After a rest period of several weeks to permit the induction of sensitivity, another application of the chemical is made and the reaction is compared to the initial reaction. These tests are excellent for predicting strong sensitizers, but less reliable indicators of weak sensitizers. This poses a problem for identifying weak sensitizers that may be used by large numbers of people.

Photosensitization

Photosensitization is defined as an abnormal adverse reaction of the skin to ultraviolet (UV) and/or visible light. Foreign chemicals may induce such reactions in several ways. The two most important are *phototoxicity* and *photoallergy*.

Phototoxicity is a chemically-induced increased reactivity of a tissue after exposure to UV and/or visible radiation through a non-immunologic mechanism. Phototoxic responses are dependent on the nature and concentration of the chemical in the skin and the amount of light exposure. Phototoxic reactions may occur after contact, ingestion, or injection of causal agents. The skin and eyes are the major organs affected. Pathologic changes may include swelling, redness, and blistering. Hyperpigmentation may also occur after a reaction.

Phototoxic reactions can be grouped into those that require oxygen (photodynamic reactions) and those that do not (nonphotodynamic reactions). In photodynamic reactions, light of the appropriate wavelength reacts with the foreign molecule and excites it. The excited molecule then reacts with oxygen to produce highly reactive oxygen free radicals which can damage tissue. In nonphotodynamic reactions, the chemical is excited by light, but then reacts directly with cellular proteins, lipids, and DNA. Phototoxic agents include furocoumarins such as 8-methoxypsoralen which is used to treat psoriasis, polyaromatic hydrocarbons (anthracene, acridine, and phenanthrene), tetracyclines, and phenothiazides.

Photoallergic reactions result from cell mediated immune reactions similar to allergic contact dermatitis. Photoallergic reactions cause dermatitis which can also spread to areas of the skin not exposed to the light. The vast majority of these reactions result from topically administered agents interacting with UV light.

The role of light in photoallergy is believed to be the conversion of a chemical to an allergic form. Two mechanisms have been proposed. Radiation may chemically change a photosensitizer into an antigenic form. An example of this would be sulfanilamide which is converted by UV light to the potent allergen *p*-hydroxylaminobenzene sulfonamide. Alternatively, the light-excited photosensitizer may react with protein producing an antigenic photosensitizer-protein product. Some reported photoallergens are halogenated salicylanilides, sulfonamides, coumarin derivatives, sunscreen components (glycerol *p*-aminobenzoic acid), and several plant products. In general, photoallergens have little phototoxic activity.

Chemical Acne and Chloracne

A number of agents produce conditions similar to those seen in acne vulgaris, the typical form of acne. These include greases and oils, coal tar pitch, creosote, and a number of cosmetic preparations. Systemically administered drugs including iodides, bromides, and isoniazid can also cause acne.

Chloracne is a more specific and severe form of acne. It is characterized by small, strawberry colored cysts and comedomes (blackheads). Inflammatory pustules and abscesses may also occur. The most sensitive areas of the skin are below and to the outer side of the eye and behind the ear. The eruption may be limited to these areas or may involve nonexposed areas, especially the scrotum. Histologic changes begin with keratinization of sebaceous gland duct epithelium and outer root sheaths of hair follicles. The sebaceous gland and follicle sheaths become replaced by a keratinous cyst. Chloracne may continue well after initial exposure to the inciting agent because of the release of the chemical from body stores.

Agents that cause chloracne include polyhalogenated dibenzofurans, polychlorinated dibenzodioxins (including 2,3,7,8-tetrachlorodibenzo-*p*-dioxin or TCDD), polychloronapthalenes, polyhalogenated biphenyls (PCBs and PBBs), and polychlorinated azoxybenzenes. Many of these substances are byproducts during the manufacture of other polychlorinated substances. TCDD is the most potent inducer of chloracne. The development of chloracne following exposure to PCBs and PBBs is probably due to contaminants such as tri- and tetrachlorodibenzofurans (TCDFs) rather than the PCBs or PBBs themselves.

Other Toxic Cutaneous Reactions

A wide variety of other toxic skin reactions may occur in addition to those already discussed. These are best described by morphologic criteria. *Physical dermatitis* is produced by fiberglass and other fibers and results in an intense pruritic (itching) reaction and pinpoint-sized reddened papules. The development of this condition is directly related to the fiber diameter (which must be greater than 4.5 μm) and inversely related to fiber length. *Chemical burns* are caused by corrosive substances and result in severe ulceration of the skin. Acids produce severe burns with a dry crust. Alkali produce softer, but extremely painful burns. Phenolics cause anesthetic effects in the skin so that pain may be absent or less severe after a short period of time. Some chemicals such as nitrogen mustard and organotins produce delayed burns in the skin. *Urticarial reactions* (wheal-and-flare reactions) are produced within 30 to 60 min of exposure to a variety of agents. They are caused by a variety of chemicals found in plants (nettles), animals (caterpillars and jellyfish), and other sources. These reactions are thought to be due to the release of histamine and other substances from immune cells and are characterized by rapidly occurring intense burning, itching, and redness. *Cutaneous granulomas* usually appear as slightly reddened, more or less flesh-colored papules that may be clustered and associated with inflammatory changes. They are generally localized to areas of foreign substance contact. They result from mononuclear cells attempting to "wall off" poorly soluble foreign body and antigenic substances in the dermis. These include talc and silica and certain metals such as beryllium, zirconium, and chromium salts. *Hair damage and loss* may occur through direct effects on the external hair or through damage to the hair producing cells of the hair follicle. Alkali, thioglycolates, and oxidizing agents such as peroxides and perborates can dissolve hair keratin and cause softening, matting, and increased fragility of the hair. Growth inhibitory agents such as alkylating agents, antimetabolites, and colchicine may impede the growth of follicular cells and the production of hair keratin. In addition, dyes (indigo), metals (copper and cobalt), and acids (picric acid) may change the color of the hair upon contact. *Hypopigmentation* of the skin can be caused by a number of agents especially phenols and catechols that inhibit melanin production. *Hyperpigmentation* is a secondary response to phototoxicity as discussed above. Coal tar pitch, psoralens, heavy metals (silver, arsenic, mercury, and bismuth), acridines, phenothiazines, and tetracyclines are examples of photosensitizers that induce hyperpigmentation. *Skin cancers* include basal cell carcinoma, melanoma, and metastatic tumors that arise from cancers of other organs. Important human skin carcinogens are UV and ionizing radiation, polyaromatic hydrocarbons (PAHs), arsenic, and combined exposure to psoralens and UV light.

REFERENCES

Adams, R. M., *Occupational Skin Diseases*, Grune & Stratton, Inc., New York, 1983.

Fisher, A. A., *Contact Dermatitis*, Lea & Febiger, Philadelphia, 1986.

Kimber, I. and Maurer, T., *Toxicology of Contact Hypersensitivity*, Taylor & Francis, Bristol, Pennsylvania, 1996.

Klaassen, C. D. (ed), *Cassarett and Doull's Toxicology: The Basic Science of Poisons*, 5th edition, Pergamon Press, New York, 1995.

Maibach, H. I., *Occupational and Industrial Dermatology*, Year Book Medical Publishers, Chicago, 1987.

Marlzulli, F. N. and Maibach, H. I., *Dermatology*, 5th edition, Taylor & Francis, Bristol, Pennsylvania, 1996.

11

Toxic Responses of the Nervous System

Joseph C. Siglin

CONTENTS

1-56670-239-9/97/$0.00+$.50
© 1997 by CRC Press, Inc.

INTRODUCTION

The nervous system is an exceedingly complex network of specialized organs and tissues that provide sensory, integrative, and motor functions essential for normal homeostasis. Over the years, many of the structural and functional aspects of the nervous system have been explored and elucidated. However, with regard to neurotoxicity, relatively little is known concerning the specific mechanisms of action of many neuropoisons. Moreover, the potential neurotoxic effects of numerous chemicals has not received adequate attention in laboratory studies designed for human risk assessment. The purpose of this chapter is to briefly review some basic aspects of normal neuronal structure and function, and to describe the target sites and clinical symptoms of neuronal dysfunction produced by a variety of occupationally encountered neurotoxic agents. Specific mechanisms of neurotoxicity will be discussed (when known) and methodologies for detecting and testing potential neurotoxicants will be presented.

ORGANIZATION OF THE NERVOUS SYSTEM

The nervous system is comprised of the brain, the spinal cord, and an extensive network of nerves and sensory organs. The two principal divisions of the nervous system are the *Central Nervous System (CNS)* and the *Peripheral Nervous System (PNS)*. The CNS serves as the control center for the entire system and consists of the brain and spinal cord. The PNS consists of countless nerve processes that connect the CNS with various receptors, muscles, and glands. The PNS may be divided into the *afferent*, or sensory system, and the *efferent*, or motor system. The afferent system conveys information from receptors to the CNS, while the efferent system conveys information from the CNS to muscles and glands. The efferent system may be further divided into the *somatic nervous system*, which conveys information from the CNS to skeletal muscles, and the *autonomic nervous system*, which conveys information from the CNS to smooth muscle, cardiac muscle, and glands. Finally, the viscera of the body receive nerve fibers from two divisions of the autonomic nervous system known as the *sympathetic* and *parasympathetic nervous systems*. In general, fibers of the sympathetic nervous system stimulate organ activity, while fibers from the parasympathetic system decrease or restrict organ activity.

NEURONAL STRUCTURE AND FUNCTION

The fundamental cell of the nervous system is the *neuron*. The neuron is highly specialized for conduction of nerve impulses and is responsible for all special functions attributed to the nervous system. Each neuron consists of a cell body known as the *perikaryon* and specialized extensions known as *axons* and *dendrites* along which electrical impulses (action potentials) travel. These impulses are generated as a result of rapid changes in the flow of sodium and potassium ions through channels in the cell membrane. The external sheath of the neuron is composed of *myelin*, a phospholipid which acts as an insulating material to further accelerate electrical impulses generated by the neuron. In general, electrical impulses travel away from the perikaryon along axons and interact with dendrites of opposing neurons. Although all neurons conform to the general description given above, there exists considerable differences in neuronal structure and function at various locations within the body.

Impulses are conducted from one neuron to another across a junction referred to as the *synapse*. At the synapse, the axonal terminals of presynaptic neurons terminate in bulb-like structures referred to as *synaptic end bulbs*. These end bulbs commonly synapse with the dendrites or cell bodies of one or several postsynaptic neurons. Whether an impulse is conducted across a synapse depends on the presence of chemical messengers which are collectively referred to as *neurotransmitters*. The neurotransmitters are synthesized by the neuron, transported to the synaptic end bulbs, and stored in small membrane bound sacs called synaptic vesicles. Upon stimulation of the end bulbs, neurotransmitter molecules are released into the synaptic cleft where they subsequently interact with receptors located on the postsynaptic plasma membrane. This interaction may result in initiation or inhibition of impulse conduction in the opposing neuron. The specific outcome depends on the chemical nature of the neurotransmitter and its interaction with the receptors of the postsynaptic neuron. In either case, impulse conduction occurs in only one direction, from the presynaptic neuron to the opposing postsynaptic neuron.

Nerve impulses are conducted to muscle cells across an area of contact referred to as the *motor end plate*. Similarly, nerve impulses are conducted to glandular cells across an area of contact known as the *neuroglandular junction*. The principal mechanisms involved in muscle and gland stimulation are similar to those described above for neuron-to-neuron impulse conduction and again involve the release of neurotransmitter molecules from presynaptic end bulbs.

The neurotransmitters may be broadly divided into two categories: the classical neurotransmitters which include acetylcholine, norepinephrine, serotonin, and dopamine, and the neuropeptides which include endorphin, enkephalin, vasopressin, and substance P. Unlike the classical neurotransmitters which are somewhat localized in their activity, the neuropeptides may act over long distances, traveling through blood vessels to interact with receptors on distant neurons and other tissues. Like the structural features of the nervous system itself, neurotransmitters are subject to alterations which could potentially manifest in abnormal neurological symptoms.

NEUROTOXICITY

Broadly defined, a neurotoxic substance is any substance capable of producing an adverse effect on the structural or functional components of the nervous system. The nervous system is particularly vulnerable to toxic insult since nervous tissue has a limited or negligible capacity for regeneration. The effects of neurotoxins may be immediate or delayed in onset and the ensuing alteration may be permanent or reversible. For many neurotoxins, adverse symptoms may be evident after only a single exposure. For still other neurotoxins, effects may manifest only after repeated exposure for weeks or years. Like other toxins, exposure to neurotoxic agents may occur by various routes (e.g., absorption through the skin, ingestion, or inhalation).

Neurotoxic substances are diverse in their chemical structure, source, and mode of action. They include naturally occurring elements such as mercury and lead, biological compounds such as botulinum toxin and tetrodotoxin, and synthetic compounds such as pesticides and industrial solvents. A number of therapeutic drugs and drugs of abuse also have neurotoxic potential. Neurotoxic substances may produce a spectrum of adverse neurological effects. These include motor effects such as convulsions, tremors, lack of coordination, paralysis, and weakness; sensory effects such as equilibrium changes and visual, tactile and auditory disorders; cognitive effects such as confusion, memory deficits, and speech and learning impairments; mood and personality effects such as depression, excitability, irritability, restlessness, delirium, and hallucinations; and generalized effects such as loss of appetite, fatigue, and overall depression of neuronal activity. In any case, substances that affect sensory, motor, learning, or memory skills are deemed neurotoxic, even if the specific mechanism(s) for these effects are not known.

Neurotoxins may elicit their effects at the molecular or cellular level. For example, the neurotoxin may interfere with neurotransmitter synthesis leading to reduced production and subsequent neuronal dysfunction. Alternatively, a neurotoxin may alter ionic flow across the nerve cell membrane leading to inhibition of impulse formation and transmission. Neurotoxins may also selectively damage the neuronal body (*neuropathy*), axons (*axonopathy*), or myelin (*myelinopathy*). These latter types of changes often result in slow degeneration of the cell body or axon and lead to permanent neuronal damage.

As compared to adults, developing fetuses and children are more vulnerable to the effects of certain neurotoxins. This increased vulnerability is attributable to several factors. First, since the nervous system is in an active growing stage in developing fetuses and children, it is often more readily perturbed by certain neurotoxic agents. Secondly, in fetuses and children, the so called "*blood-brain barrier*" is not yet completely formed. In adults, the blood-brain barrier is composed of a layer of tightly juxtaposed cells in blood vessel walls of the brain which selectively restrict entry of molecules to those necessary for metabolic function. Finally, in fetuses and children, metabolic pathways which detoxify potentially damaging xenobiotics such as neurotoxins are not fully developed. Thus, exposure to neurotoxic agents which the adult may tolerate or overcome may permanently damage the more sensitive nervous system of the developing fetus or child.

The elderly represent another group possessing an increased sensitivity to neurotoxic damage. In this case, the age-related decline of the nervous system limits its ability to compensate for neurotoxic effects. The decreased hepatic and renal functions of the elderly also contribute to the enhanced susceptibility of this group to the toxic effects of various xenobiotics, including neuropoisons.

There are a number of difficulties in evaluating the potential risk of neurotoxic drugs and chemicals. For example, it is possible for a substance to be safe and beneficial at one concentration but neurotoxic at another. Such is the case with vitamins A and B6 which are required in the diet in trace amounts, but which may cause neurotoxicity in large doses. Similarly, antipsychotic drugs have been useful in treating schizophrenia, yet chronic use of some antipsychotics may give rise to severe, incapacitating neurological effects. The potential additive effects of toxic substances may further complicate efforts to evaluate neurotoxicity. Such situations may occur following long-term exposure to complex mixtures of chemicals in the workplace.

CLASSES OF NEUROTOXIC SUBSTANCES

Neurotoxic substances are most conveniently categorized according to the types of structural or functional changes they induce. Common sites of toxic action for neurotoxins include the neuronal cell membrane, neuronal structures, glial cells and myelin, and the neurotransmitter system.

Substances Acting on the Neuronal Cell Membrane

The neuronal cell membrane contains a complex system of pumps, receptors, and channels through which ions travel into and out of the cell. All of these components serve as potential targets for neurotoxicants. Blockage of the ion channels may initially cause numbness, followed by an inability to coordinate muscular movements such as speaking and swallowing. In severe situations, respiratory paralysis may ensue. Examples of such channel blocking substances include naturally occurring neurotoxins such as tetrodotoxin and saxitoxin. Tetrodotoxin is found in the liver of the puffer fish, while saxitoxin is produced by certain marine alga and is responsible for paralytic shellfish poisoning. Toxic substances may also increase ion flow across the neuronal membrane, leading to many of the same symptoms produced by channel blockers. Examples of these latter agents include the pyrethroid pesticides and the well known organochlorine pesticide, DDT.

Substances Acting on Neuronal Structures

Degeneration of the CNS may be produced by substances such as mercury and lead. Exposure to organic mercury may cause degeneration of neurons in the cerebellum leading to tremors, difficulty in walking, and visual impairment. Similarly, exposure to lead may adversely affect the cortex of the developing brain leading to mental retardation. Children are particularly susceptible to the effects of these compounds owing to the actively developing state of the immature brain.

As compared to the CNS, the PNS is often more vulnerable to neurotoxic effects since it lacks the protective features of the blood-brain barrier. One of the most frequently encountered effects in the PNS consists of degeneration of the neuronal axons (*axonopathy*). Many chemicals and drugs are known to induce axonal degeneration, without directly affecting the cell body. Often, the effects of such agents are manifested only after repeated, prolonged exposure conditions. The mechanisms responsible for producing axonopathy are not well understood; however, it appears that the axonal degeneration may involve block-age of transport of vital substances between the cell body and axon. Although the cell body is not directly involved, the degeneration often proceeds toward the cell body, leading to the pathological condition termed *central-peripheral distal axonopathy (CPDA)*. Industrial chemicals that cause CPDA include carbon disulfide, hexane, acrylamide and Lucel-7. Drugs that produce CPDA include thalidomide and the anticancer agent, vincristine. Other agents that may affect the nervous system in this manner include alcohol (usually in abuse situations), and some organophosphorus pesticides. Clinical symptoms of CPDA often involve loss of sensation in the extremities and muscle weakness. These effects may worsen as the degeneration ascends the limbs. On the other hand, recovery is often possible following termination of the exposure condition.

Substances Acting on Glial Cells and Myelin

A number of neurotoxic substances may induce degeneration of glial cells and the myelin which they produce. These substances are diverse in both their chemical structure and mechanisms of action. For example, Diphtheria toxin interferes with the cell bodies of the myelin producing glial cells, while hexachlorophene, a commonly used antibacterial agent in soaps and antiseptic solutions, interferes with the mitochondria of glial cells producing adverse, energy deficient states within the cell.

Substances Acting on the Neurotransmitter System

The neurotransmitter system of the neuron represents yet another target for certain neurotoxic substances. A classic example of neurotoxicity due to disruption of the neurotransmitter system involves inhibition of acetylcholinesterase by various organophosphorus insecticides. During normal impulse conduction, the neu-rotransmitter, acetylcholine, is released from synaptic end bulbs of an axon into the synapse. The neurotransmitter molecules then interact with receptors on the dendrite of an opposing neuron to continue impulse conduction. An endogenous enzyme, acetylcholinesterase, then inactivates the acetylcholine molecules. How-ever, when present in the synaptic region, organophosphorus compounds bind with the endogenous acetylcholinesterase molecules rendering the enzyme inef-fective. Consequently, the acetylcholine transmitter accumulates in the synapse and inappropriately continues impulse conduction. The signs and symptoms of organophosphate poisoning are predictable based on this mechanism of action and include increased salivation, lacrimation, and sweating; ataxia, tremors, muscle fasciculations, and convulsions; constriction of the pupils; bronchial constriction; and increased peristalsis with consequent development of nausea, vomiting, abdominal cramps, and diarrhea. In fatal organophosphorus poisoning, death is due to asphyxiation resulting from respiratory failure. However, in most acute exposures, full recovery occurs following cessation of the exposure.

Drugs of abuse such as cocaine, LSD, morphine, and heroin also interfere with neurotransmitter function. Cocaine affects the reuptake of the neurotransmitters, norepinephrine and dopamine, causing hyperactivity, aggression, paranoia, high blood pressure, and abnormal heart rhythms. LSD, a potent hallucinogen, appears to interfere with the neurotransmitter serotonin, while the opium-related drugs, morphine and heroin, act on receptors in the brain which are responsible for the feelings of sedation and euphoria.

OCCUPATIONAL NEUROTOXICANTS

Lead

Lead exists in both the organic and inorganic forms. However, organic lead degrades quickly in both the atmosphere and body and it constitutes only a small portion of the total lead to which the population is exposed. Inorganic lead, on the other hand, remains a significant problem to both humans and the environment. Sources of exposure to inorganic lead include water, food, soil, lead-based paint, leaded gasoline, and industrial emissions. Lead exposure is often contingent on location. For example, in some areas, exposure is dependent on proximity to lead producing industries such as smelters, whereas in older cities, the principal source of exposure is often leaded paint. As compared to adults, children typically ingest and inhale more lead per unit of body weight and are more susceptible to its adverse effects. Because lead imparts a sweet taste to paint, it is appealing for children to ingest lead containing paint chips.

Both the absorption and retention of lead is higher in children than adults. The increased retention of lead in children is attributable to its active disposition and subsequent storage in growing bones. Since children have less overall bone storage mass, lead may remain in the bloodstream, free to exert its effects on the various organs and tissues of the body. As mentioned previously, children are more vulnerable to the effects of lead due to the immature nature of their nervous system, particularly their blood-brain barrier which is not fully developed. Although lead is now rarely used in paint products, past use of lead-containing paint continues to present a significant source of lead exposure to children, particularly in impoverished urban areas.

For adults, the workplace represents the major source of potential lead exposure. In the United States, The National Institute of Occupational Safety and Health (NIOSH) has identified more than 100 occupations that potentially increase adult exposure to inorganic lead. For adults that are occupationally at risk, and for children older than six years, the major sources of lead exposure are contaminated food and water.

Routes, Levels, and Systemic Effects of Lead Exposure

The three primary routes by which lead may enter the body are inhalation, ingestion, and absorption through the skin. Dermal absorption is, however, only significant for organic forms of lead. Intake of lead through inhalation is dependent on particle size and solubility in body fluids. Gastrointestinal absorption is dependent on several factors, primarily age and nutritional status. In children, nutritional deficiencies often contribute to higher levels of lead absorption.

Because lead is stored in circulating blood, soft tissue, and bone, the concentration of lead in blood alone is not an accurate indicator of the total body burden. However, for the most part, neurological effects in adults have not been observed below the blood lead level of 40 μg/dL. In contrast, adverse neurological effects (neurobehavioral dysfunction) have been noted in children at much lower lead blood levels. Currently, the U.S. Environmental Protection Agency considers 10 to 15 μg/dL as the maximum acceptable lead blood level for children. Exposure to lead may cause numerous adverse health effects including brain injury, nerve dysfunction, decreased I.Q., chronic neuropathy, anemia, reduced heme synthesis, enzyme inhibition, elevated blood pressure, and heart injury. In children, brain damage may range in severity from impaired muscular coordination to stupor, coma, and convulsions at high levels.

Neurotoxic Pesticides

Pesticides are defined as any substance or mixture intended for preventing, destroying, repelling, or mitigating insects, rodents, nematodes, fungi, weeds, or other forms of life declared to be pests. Human exposure to pesticides may occur through contaminated drinking water, residues in food, and through production and use of pesticidal products. In the United States, approximately one billion pounds of pesticides are used annually. Although a number of pesticide-related health problems have been documented, accurate estimates of pesticide-related health problems are difficult since relatively few exposure incidents are actually reported. For those cases that are reported, the most common offending agents have been neurotoxic pesticides. Examples of pesticides known to produce neurotoxic effects are presented in Table 11.1. Some classes of neurotoxic pesticides are described below.

Cholinesterase-Inhibiting Insecticides

Organophosphorus (OP) and carbamate insecticides are the most common causes of agricultural poisonings. As discussed previously, both classes of insecticides can inhibit acetylcholinesterase, the enzyme responsible for breaking down the neurotransmitter, acetylcholine. The subsequent build-up of the neurotransmitter causes central nervous system dysfunction. Common symptoms of acute poisoning include hyperactivity, neuromuscular paralysis, visual and breathing difficulties, vomiting, diarrhea, weakness, and possible convul-

TABLE 11.1   Examples of Neurotoxic Pesticides

Organophosphorus Insecticides

tetraethyl pyrophosphate
fensulfothion
mevinphos
ethyl parathion
azinphos-methyl
chlorfenvinphos
dichlorvos
trichlorfon
dimethoate
diazinon
malathion
ronnel

Carbamate Insecticides

aldicarb
carbofuran
formetanate
propoxur
carbaryl

Organochlorine Insecticides

endrin
aldrin
dieldrin
lindane
DDT
heptachlor
chlordane
mirex
methoxychlor

sions, coma, and death. The onset and duration of symptoms is dependent on insecticide dose, route of exposure, and the inherent toxicity of the particular compound. Following acute exposure to carbamate insecticides, the inhibition of acetylcholinesterase is usually readily and rapidly reversible. In contrast, inhibition of the neurotransmitter is more persistent with OP insecticides, and thus, recovery is usually much slower. Delayed effects such as irritability, depression, mood swings, lethargy, and short-term memory loss may occur following an episode of acute OP or carbamate poisoning. It is often unclear whether these changes are the actual result of the acute poisoning, or due to prolonged low-level exposure to the pesticide. Symptoms of acute high-level exposure to OP and carbamate insecticides are summarized in Table 11.2.

**TABLE 11.2   Common Symptoms of Acute High-Level Exposure to
Organophosphorus and Carbamate Insecticides**

Normal Physiologic Response[a]	Altered Physiologic Response[b]
Activation of salivation, sweating, and tearing	Increased salivation, sweating, and tearing
Constriction of bronchi	Tightness in chest, coughing, and breathing difficulty
Constriction of pupils	Pinpoint pupils and blurred vision, increase in blood pressure
Increased spasms in digestive tract	Stomach cramps, nausea, vomiting, and diarrhea
Increased spasms in urinary tract	Frequent urination and/or incontinence
Activation of skeletal muscles	Twitching, restlessness, impaired coordination, muscle weakness, paralysis, asphyxiation, and death
Altered brain function	Headache, giddiness, anxiety, lethargy, confusion, CNS depression, and coma

[a] Function of nervous system when stimulated by acetylcholine.

[b] Effect of excessive stimulation by acetylcholine which occurs when acetylcholinesterase is inhibited.

Some OP insecticides can also produce delayed, persistent neuropathy by damaging certain neurons in the spinal cord and peripheral nervous system. Initial symptoms of this neuropathy usually include cramps in the calves and tingling in the feet. These symptoms are followed by flaccidity of the legs and varying degrees of sensory disturbance. The rate and extent of recovery from these effects vary considerably.

Organochlorine Insecticides

Organochlorine insecticides are chlorinated hydrocarbon compounds that generally act as nervous system stimulants. Although these compounds are less acutely toxic than OP or carbamate insecticides, they possess a greater potential for chronic toxicity due to their accumulation in the environment and the body. Probably the best known and studied organochlorine insecticide is chlorophenothane, better known as DDT. Before it was banned by the EPA in 1972, DDT was used extensively in agriculture and against mosquitoes and other insects which transmit human diseases. The EPA has also banned or severely restricted the use of other organochlorine compounds such as aldrin, dieldrin, toxaphene, mirex, heptachlor, and chlordane, following recognition of their accumulation in the environment, accumulation in human and animal tissues, and adverse affects on some wildlife.

Organochlorine compounds are readily absorbed by inhalation, ingestion, and through the skin. Following absorption, these compounds are generally distributed to fatty tissue, liver, and the central nervous system. Elimination of these compounds from the body is usually a very slow process. This is particularly true following chronic exposure which results in extensive accumulation of the compounds in fatty tissues. Acute exposure can produce nervous system excitability, apprehension, headache, dizziness, confusion, weakness, tremors, and possible convulsions, coma, and death. Children are particularly vulnerable to brain and nerve tissue damage from organochlorine pesticides and may suffer persistent behavioral and learning disabilities.

Fumigants

Fumigants are gases used to kill insects, insect eggs, and microorganisms. One of the most widely used fumigants in the U.S. is methyl bromide, a colorless gas with a faint, somewhat agreeable odor. This pesticide has caused severe neurotoxic effects and death in fumigators, applicators, and pest control workers. Acute exposure can result in visual and speech disturbances, delirium, and convulsions. Both acute and chronic poisoning may be followed by prolonged, sometimes permanent, brain damage which may manifest as marked personality changes and perceptual difficulties. In some cases, chronic exposure may result in progressive peripheral neuropathy, with loss of motor control, numbness, and weakness.

Pyrethroids

The pyrethroids are a group of insecticides which are highly toxic to insects, but less toxic to mammals. The mechanism of action of pyrethroids involves alteration of the flow of sodium ions through the nerve cell membrane, resulting in repeated firing of the nerve cell. Because pyrethroids appear to be less acutely toxic than other insecticides, their development and use may be expected to increase in the coming years.

Organic Solvents

Organic solvents are simple organic liquids which are used in a variety of industrial processes and commercial products such as paints, paint removers, and varnishes; adhesives, glues and coatings; degreasing and cleaning agents; dyes and inks; floor and shoe polishes; agricultural products; pharmaceuticals; and fuels. Acute exposure to many organic solvents can produce feelings of inebriation and loss of manual dexterity, coordination, and balance. Chronic exposure to some organic solvents can produce fatigue, irritability, loss of memory, and changes in mood and personality. In some cases, actual structural changes in the nervous system may occur. Common classes and examples of various organic solvents are presented in Table 11.3.

Uptake, Distribution, and Elimination of Solvents

Solvents may enter the body by inhalation, ingestion, or dermal absorption. However, because organic solvents are volatile, the inhalation route constitutes the major pathway of accidental solvent exposure. The amount of solvent

TABLE 11.3   Classes of Organic Solvents

Chemical Class	Example
Aliphatic hydrocarbons	n-Hexane
Cyclic hydrocarbons	Cyclohexane
Nitrohydrocarbons	Ethyl nitrate
Aromatic hydrocarbons	Benzene
Halogenated hydrocarbons	Carbon tetrachloride
Esters	Ethyl acetate
Ketones	Acetone
Aldehydes	Acetaldehyde
Alcohols	Methyl alcohol
Ethers	Ethyl ether
Glycols	Ethylene glycol

entering the body via the inhalation route depends on its concentration in the air, its solubility in blood, and the ventilation rate and depth during exposure. Some solvents tend to be unequally distributed among the organs of the body due to differences in tissue solubility and blood perfusion. Typically, the brain achieves high solvent levels quickly due to its high fat content and rich blood supply. Solvents are eliminated from the body through exhalation or metabolism. As with other xenobiotics, the metabolic products of solvents may be more toxic than the parent compound itself.

Neurological and Behavioral Effects

All solvents are fat soluble and will produce central nervous system effects at some level of exposure. Short-term exposure at low toxicity levels may produce mucous membrane irritation, tearing, nasal irritation, headache, and nausea. At higher levels, there may be initial euphoria followed by confusion, motor incoordination, ataxia, unconsciousness, and even death. Because solvents can potentially impair work performance and the ability to avoid hazards, use and exposure to solvents in the work place is closely regulated by the NIOSH. Toxicity studies have revealed a number of specific effects associated with particular solvents or classes of solvents. For example, *hydrocarbon neuropathy* may result from chronic exposure to hexane, methyl-*n*-butyl ketone, and related solvents. This disorder is characterized by numbness in the hands and feet which may progress to generalized muscle weakness and lack of coordination. Other examples of solvent-specific neurological effects include high-frequency hearing loss due to high levels of xylene or toluene; seizures and convulsions due to acute exposure to methylcyclopentane or methylcyclohexane; damaged facial nerves and facial numbness due to trichloroethylene (or its contaminants); and impaired perceptual speed, accuracy, memory, and cognitive performance due to styrene. Still other solvents may produce emotional disorders, such as sleep disturbances, nightmares, and/or insomnia, while severe brain injuries (chronic encephalopathies) may be produced following repeated, prolonged exposure to some solvents (i.e., occupational exposure or exposure from deliberate self-administration of solvents). The encephalopathy is characterized by wasting of the brain matter and dilation of the fluid-filled cavities in the brain with resulting motor dysfunction and impaired mental functions.

CONTROLLING WORKER EXPOSURE TO SOLVENTS

It has been estimated that 9.8 million workers come into contact with solvents or solvent mixtures daily through skin contact or inhalation. Thus, worker safety is of paramount importance in avoiding or minimizing worker exposure. Methods utilized for controlling worker exposure include worker isolation, appropriate engineering controls, and personal protective equipment. Of these methods, suitable engineering controls represent the primary means of preventing worker contamination. Closed system operations are, of course, the most effective method of minimizing worker exposure. However, when a closed system is not practical, negative air pressure hoods and fans should be utilized to direct solvent vapors away from the worker. In addition, respirators, face shields, safety goggles, solvent resistant gloves, and suits should also be utilized to avoid exposure.

The Occupational Safety and Health Act (OSHA) of 1970 was enacted to protect workers from occupational hazards in the workplace (CFR 1987 ed. 1900–1910). According to the law, employers are required to maintain a workplace "free of recognized hazards" with work environments that meet the regulations and standards set forth by the law. In administering the law, OSHA conducts inspections, determines compliance with the standards, and initiates actions against employers who are not in compliance. In 1989, revised permissible exposure limits (PELs) were published by OSHA for 428 toxic substances, including a number of organic solvents. Former and revised OSHA PELs are presented for several selected solvents in Table 11.4.

Educational programs geared toward hazard identification and protection are, of course, an important component of any program designed to heighten worker safety. In addition, NIOSH recommends that appropriate medical surveillance programs be implemented to assess potential acute and chronic exposure situations in the workplace. Both the attending physician and the worker should be provided with information concerning the adverse health effects of the solvent and an estimate of the worker's potential for exposure. The results of the periodic medical examinations should be correlated with results of workplace sampling to provide the best overall assessment of worker exposure and safety.

TESTING OF NEUROTOXIC SUBSTANCES

There are a number of approaches to testing potential neurotoxic substances, including whole animal studies, cell culture tests, and tests on human subjects.

TABLE 11.4   Permissible Exposure Limits (PELs) for Some Selected Solvents

Solvent	Former PEL[a] (ppm)	New PEL[a] (ppm)
Toluene	200	100
Xylene	100	100
Acetone	1000	750
Styrene	100	50
Tetrachloroethylene	100	25
Furfuryl alcohol	50	10
Ethylene dichloride	50	1
Benzene	10	10
Carbon disulfide	20	4
Trichloroethylene	100	50
Chloroform	50	2

[a] PELs are set by the Occupational Safety and Health Administration (OSHA).

Of course, human subjects provide the most reliable information concerning the potential adverse health effects of toxic substances; however, in addition to being difficult and dangerous, this approach is generally viewed as unethical. Thus, cell culture (*in vitro*) and whole animal (*in vivo*) studies currently offer the best alternative means for studying potential neurotoxic agents.

Animal Tests

From a biological stand point, animals resemble humans in many ways. However, differences exist in many biochemical and physiological processes between animals and humans, and it is often difficult to extrapolate results of animal studies to humans. These differences necessitate careful consideration of the experimental design, animal choice, dosing regimen, route of exposure, and the extent and duration of treatment. These factors are briefly discussed below.

Experimental Design

Prior to undertaking any animal toxicity study, the investigator must first clearly define the goals and objectives of the research and understand how the resulting data may be used. Experimental variables which should be specifically considered in neurotoxicity testing include evaluation of the nervous system for physical or physiological alterations, specific characterization of any resulting neurological alterations, and identification of any primary targets of the test substance. If there are no existing toxicology data for the test substance, preliminary screening may be undertaken using fewer animals to investigate the consequences of acute or short-term exposure prior to initiating a definitive, multi-level study. The ensuing definitive study should evaluate a variety of experimental parameters to accurately assess potential functional, morphological, or physiological alterations to the nervous system. The experimental design should routinely employ multiple groups and graded dosage levels of the test substance to accurately assess potential dose-response relationships, and hopefully, to establish a no-observed effect level (NOEL). All observed effects should be closely followed with respect to onset and duration. In addition, an appropriate untreated, vehicle-treated, or sham-treated control group should be included in the experimental design for the purpose of comparison. The investigator should also consider including appropriate positive controls for the purpose of test validation, particularly if adequate historical data are not available, or if the testing methodologies have not been adequately validated. Finally, the investigator should consider including additional (satellite) groups of animals in which dosing is discontinued at some point to allow evaluation of the potential reversibility of effects.

Animal Model

The animal model selected for testing is of critical importance in all toxicological investigations. For evaluation of organophosphorus compounds, hens have been utilized owing to their established sensitivity to this class of neurotoxic compounds. However, for most other neurotoxicity screening studies, the laboratory rat has served as the species of choice, because: (1) it has been shown to be sensitive to a variety of neurotoxic agents, (2) an extensive data base of toxicological information exists for various strains of laboratory rats, (3) the rat is a reasonable alternative to larger mammals for toxicity testing for both economical and practical reasons, and (4) genetically similar inbred strains of rats may be obtained from reliable, USDA regulated animal suppliers.

Other factors which should be taken into consideration in selection of animals for neurotoxicity studies are animal age and sex. Since some agents may produce a greater effect in only one sex, both males and females should be routinely tested in neurotoxicity evaluations, and in toxicological tests in general. Currently, the EPA requires the use of both males and females for neurotoxicity evaluations. Animal age may also dramatically affect response to certain neurotoxic agents. For example, the normal loss of nerve cells with increasing age and age-related decline of nervous system function may predispose older animals to the toxic effects of some neurotoxins. Similarly, the developing nervous system of younger animals may also be more vulnerable to the adverse effects of various neurotoxins. This latter situation is exemplified in humans by the increased sensitivity of children to various neurotoxic substances, including lead. In consideration of these potential influences, an ideal animal study would be one that permits longitudinal evaluation of both sexes throughout the normal stages of development (e.g., from weanlings to sexual maturity).

Dosing Regimen and Route of Exposure

The type of toxic effect elicited by a particular neurotoxin may be influenced by the level and/or frequency of dosing. In acute studies, test substances are typically administered at single high doses which often result in sudden and severe responses. In contrast, in repeated dose experiments which utilize lower individual doses, toxic manifestations may be entirely different and are typically less conspicuous and delayed in onset. For example, the primary acute effect of carbon disulfide is central nervous system depression; however, when administered repeatedly at lower exposure levels, carbon disulfide may result in peripheral neuropathy and Parkinsonism. For both acute and repeated exposure conditions, delayed neurotoxic effects may manifest following cessation of exposure. The most common routes by which neurotoxins enter the body are inhalation, oral ingestion, and dermal penetration. Therefore, these routes are those which are most commonly utilized in animal testing. Since the specific route may influence the onset, duration, and intensity of toxic responses, the actual route selected should be representative of the most likely route of human exposure.

Components of Animal Tests

In the U.S., the EPA has lead the way in developing specific test guidelines for evaluating potential neurotoxic chemicals and pesticides. When deemed necessary, a "core test battery" is required by EPA which consists of a Functional Observational Battery (FOB), an assessment of motor activity, and neuropathological evaluations. Currently, EPA's Office of Toxic Substances and Office of Pesticides Programs is considering requiring this core test battery on all new and old chemicals and pesticide products. Components of the core testing battery are briefly described in the following sections.

Functional Observational Battery

An FOB is an assortment of noninvasive tests used to evaluate sensory, motor, and autonomic nervous system functions in animals exposed to potential neurotoxic substances. The tests are usually carried out on groups of five male and five female rats. The study design usually comprises four experimental groups, representing a control and three dosage levels of the test substance. In performing the FOB, the technician first examines the animal for abnormal posture, closure of the eyelids, and presence of tremors or convulsions. Any indications of autonomic nervous system dysfunction, such as lacrimation or salivation, are then recorded. The rat is then placed on a flat surface for three minutes, during which time the number of rears are counted and the animal's gait, mobility, and level of arousal are rated. Next, the animal's response to several stimuli are rated (e.g., approach of a pencil, snap of a metal clicker, touch of a pencil on the hind quarters, and pinch of the tail). Additional FOB parameters include pupil response, righting reflex, forelimb and hindlimb grip strength, foot splay, body weight, and rectal temperature. The entire procedure usually takes about 6 to 8 min per animal. Data are subsequently summarized and statistically analyzed.

The FOB is useful since it enhances the detection of functional changes which might otherwise be overlooked. In addition, potential dose response relationships may be elucidated and the time course of various changes may be closely followed. While the FOB is useful as a screening method for potential neurotoxic substances, it is not intended to provide an overall evaluation of neurotoxicity.

Motor Activity

A number of neurotoxic substances have been found to impact motor activity in both animals and humans. Consequently, the assessment of motor activity has become an important component in standardized neurotoxicity testing. Newer automated systems have been developed to aid in the assessment of motor activity in animal experiments. These systems offer the advantages of increased accuracy, decreased subjectivity, and elimination of possible subject-observer interaction. Of the various automated systems currently available, those utilizing infrared light beams and photoreceptors appear to be most popular. As the animal moves about the test cage, beam interruptions are counted and recorded by a computer. The animal is usually allowed 30 to 60 min in the test cage environment. Data are subsequently summarized and statistically analyzed.

Neuropathology

Pathological examinations of neural tissues are performed to detect and evaluate potential histological changes in the nervous system associated with exposure to the test substance. In addition to characterizing the type, frequency, and severity of pathological lesions, the neuropathological examination helps to distinguish between physiological and structural types of adverse effects to the nervous system.

In Vitro Techniques for Neurotoxicity Testing

Various *in vitro* techniques are currently under development for use in neurotoxicological research. These include the use of primary cell cultures, tissue explant cultures, and various cell lines developed from malignant nervous tissue. While these techniques have proven useful in a number of research applications, they are not currently sufficient to replace whole animal tests for human safety evaluation and risk assessment. Nevertheless, the many advantages of *in vitro* testing and potential for reduced animal use provide strong incentives for continued development and increased utilization of *in vitro* techniques.

SUMMARY AND CONCLUSIONS

The complexity of the nervous system has made neurotoxicology one of the most demanding disciplines in toxicology. Although many neuropoisons have been characterized with respect to target sites and clinical symptoms, a great deal remains to be learned concerning the specific mechanisms of action of these agents. New testing methodologies and emerging technologies in the field of neurotoxicology will undoubtedly heighten our overall understanding of these mechanisms. However, in the mean time, increased vigilance toward identifying and testing potential neurotoxins will be needed to safeguard human populations.

REFERENCES

Hayes, A. W. (ed), *Principles and Methods of Toxicology*, 3rd edition, Raven Press, New York, 1994.

Klaassen, C. D. (ed), *Cassarett and Doull's Toxicology: The Basic Science of Poisons*, 5th edition, Pergamon Press, New York, 1995.

Office of Technology Assessment, United States Congress, Neurotoxicity: Identifying and Controlling Poisons of the Nervous System, OTA-BA-436, U.S. Government Printing Office, Washington, District of Columbia, 1990.

Wilson, C. M., Ginsberg, R., and Gordon, H. D. (eds), *Principles of Anatomy and Physiology*, 5th edition, Harper & Row, New York, 1987

12

Toxic Responses of the Female Reproductive System

Joana Chakraborty and Maureen McCorquodale

CONTENTS

INTRODUCTION

Occupational exposure to chemicals in human females may be hazardous to a wide range of reproductive processes. Altered fertility, low birth weight, spontaneous abortion, transplacental carcinogenesis, congenital malformations, mutagenesis and developmental abnormalities are among the effects on reproduction that have been recognized to result from toxic occupational exposure. Over the past several decades there has been a threefold increase in the number of women employed in the work force in the United States and many of these women have become employed in hazardous occupations including those traditionally limited to men. In addition, during recent years an increasing number of women have remained in the work place until near the end of pregnancy. Thus, many more women and their unborn children are being exposed to chemical, physical, and psychological hazards of the work place.

The Pregnancy Discrimination Act states that employers cannot discriminate on the basis of pregnancy and the possibility of harm to the mother or fetus is not grounds for "protective firing" even if it is likely that a lawsuit may later be filed against the employer for exposing the pregnant woman to an agent suspected to be teratogenic. What the employer can do, however, is limit the work exposure provided there is no "demotion" or "pay loss" associated with the change in job description.

Toxic chemicals may directly damage the female germ cells (i.e., sex cells in the ovaries), or induce adverse effects on developmental processes in the embryo and/or fetus. Injury to the germ cells in the ovary may lead to genetic abnormality which may last for generations, while embryonal and fetal damage usually lead to various deformities or mental retardation. There are at least three developmental stages during which the embryo and fetus are at risk. The first stage begins with conception and include the first 17 days following conception. During this period, the egg is fertilized to form a one-cell zygote; subsequent divisions form a ball of cells (32 to 256 cells) and these cells implant on the uterine wall. The second stage of embryonic development spans from 18 days to 55 days following conception and is characterized by folding of the embryo and organogenesis, e.g., formation of all organ systems. The third stage, which is the fetal period, spans from 56 days to term and principally involves growth and maturation of existing structures.

An acute injury during the first period of development is almost always lethal. During this stage the embryo is composed of a small number of cells and injury to these cells usually results in cell death. If a lethal dose of a chemical is given to the mother, abortion or resorption of the embryo will occur. However, if the dose is not lethal to the embryo, then recovery follows usually with no structural deformity in the fetus. If injury occurs at the second stage of development, i.e., during organogenesis then various deformities result. For example in rats, a short pulse of teratogen during midgestational period, i.e., on 10th day of pregnancy (22 days is the gestation period in rats) causes variable damage to different organs (Figure 12.1). During this stage, different organs have their own critical period of development. Therefore, this entire period is extremely sensitive. During the third stage, fully formed fetal organs are not prone to injury. Although minor malformations can be induced, the developing brain is prone to injury due to the continuous developmental process, i.e., myelination, even at the time of birth. It should be noted that development continues even after birth particularly in the skeletal, muscular and nervous systems.

Consequently, dangerous chemical exposure during the third stage of development may also lead to fetal growth retardation.

Since female reproductive toxicology involves a variety of reproductive processes, a clear understanding of each process is crucial to evaluate the reproductive risk due to chemical exposure. In this review we include four major aspects. First, the germ cell (sex cell) production in the ovary and the first stage of embryonal development will be discussed. Second, the role of cytogenetics in assessing reproductive hazards will be evaluated. Third, various

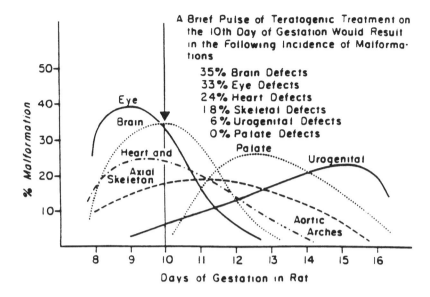

FIGURE 12.1 Group of curves representing the susceptibility of particular organs and organ systems in rat embryos to irradiation given on different days of gestation. If the agent were applied on day 10, a syndrome would result involving the organs, the curves of which are intersected by the vertical line, with percentages of incidence corresponding to the points at which the curves were crossed. Shifting the time of treatment from day 10 to another day would alter the composition of the syndrome both qualitatively and quantitatively. Modified from Wilson, J. G. *Environment and Birth Defects*. New York: Academic Press, 1973:19. From Shepard et al., *Developmental toxicology: Prenatal period*, in *Occupational and Environmental Reproductive Hazards: A Guide for Clinicians*, Williams & Wilkins, Philadelphia, PA, 1993. Chapter 4, page 45 with permission of M. Paul (ed.) and Williams & Wilkins, Baltimore, MD.

chemicals and their effects on the female reproductive system will be reviewed. Fourth, the teratogenic effects of industrial chemicals and end point evaluation for chromosome damage will be discussed.

GERM CELL PRODUCTION AND EMBRYONAL DEVELOPMENT

The female reproductive tract includes the ovary, oviduct, uterus, cervix, and vagina. Each of these organs is specialized to ensure the production of normal offspring for the propagation of the species. Although these organs are highly specialized, they are not autonomous. Proper functioning of these organs depends on the proper functioning of the neuroendocrine system as well as on the gonad (ovary) itself. The ovary has dual functions: production of female gametes, i.e., oocytes or eggs, and secretion of steroid hormones. The cortex of the adult ovary contains many follicles at various stages of development. These follicles contain an oocyte surrounded by a single layer of follicle cells.

Oogenesis

The process of germ cell production in the female is called oogenesis. In human females oogenesis starts long before birth. During the development of the ovary, germ cells undergo repeated mitoses. At 2 months gestation, the female embryo contains a large number of germ cells; some are called stem cells and others are called oogonia. Some of these oogonia enter meiosis and become primary oocytes. The primary oocytes do not have the capacity to proliferate. These cells reach the "diplotene" stage of first meiotic prophase before or shortly after birth. Shortly after the onset of the diplotene stage, oocytes embark on a prolonged "quiescent" phase, which lasts until division is resumed during oocyte maturation and shortly before ovulation. Therefore, at the time of birth, the ovaries of a human female will contain a finite number of oocytes. If these oocytes are damaged or lost, for example, due to exposure to hazardous chemicals or radiation, they cannot be replaced from stem cells and as a result the woman will be infertile for the rest of her life. If chemical or radiation injury causes chromosomal damage to the oocytes, this damage will last for a lifetime since there will be no new oocyte formation. At the time of puberty, a pool of primordial follicles grows into a pool of growing follicles. At the onset of this growth, the oocytes start to enlarge. The surrounding follicular cells divide to become several layers thick and secrete a glycoprotein material which forms an acellular layer surrounding the oocytes, called the zona pellucida. The follicular cells are called granulosa cells. Changes occur in the surrounding stromal cells which are now called theca. Some follicles grow to become preantral follicles while others undergo degeneration. This process of degeneration is prevented only if adequate follicle stimulating hormone (FSH) and luteinizing hormone (LH) are available and if the release of these gonadotropic hormones coincides with the development of FSH and LH receptors on the granulosa and thecal cells. Thecal cells divide into two distinct layers: a highly vascularized, glandular layer which is called the theca interna, and a fibrous capsulated area called the theca externa which surrounds the first layer. Granulosa cells start to secrete a fluid. Drops of this fluid coalesce to form follicular fluid within a space called the follicular antrum. At this stage, the granulosa cells surrounding the primary oocytes are called the cumulus oophorus. The primary oocytes remain suspended in the follicular fluid connected only by a thin stalk of cells to the peripheral granulosa cells. These follicles are called antral follicles. With the terminal development of antral follicles in which both theca and granulosa cells can bind LH, the follicles enter into a preovulatory growth phase. During this process the follicles enlarge leading to the expulsion of the oocytes from the follicles. This process is called ovulation. When the secondary oocytes enter into the fallopian tube at the time of ovulation, they are surrounded by layers of granulosa cells, now called corona radiata cells.

   At the time of ovulation, the oocytes undergo dramatic changes and complete the first meiotic division. This is an unusual division in which only half of the chromosomes, but almost all of the cytoplasm, goes to one cell which becomes a secondary oocyte while the remaining half of the chromosomes and a small amount of cytoplasm form the first polar body. The first polar body soon degenerates. After ovulation, the secondary oocytes begin the second meiotic division but is arrested at the metaphase until the sperm cell activates the egg at the time of fertilization. If fertilization occurs, the second meiotic division is completed. Most cytoplasm again is retained by only one cell and this cell is called the mature oocyte. The other cell receives half of the chromosomes and very little of cytoplasm. This cell is called the second polar body (Figure 12.2). After ovulation, the follicular cells undergo functional alteration and become the corpus luteum.

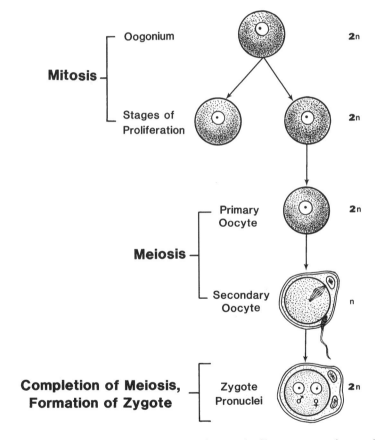

FIGURE 12.2 Diagram showing various stages of oogenesis. Chromosome numbers are indicated in the right hand side of the diagram.

It is believed that at the time of birth, the ovaries of newborn human females contain about 2 million primary oocytes. However, due to the continuous degenerating process called atresia, many of them degenerate before puberty.

At the time of puberty, only 30 to 40 thousand oocytes are present in the ovaries. Out of these, only 400 to 450 become secondary oocytes and are ovulated. Only a few of them ever reach maturity. If fertilization does not occur, the secondary oocyte degenerates in the fallopian tube (Table 12.1).

TABLE 12.1   Developmental Events

Somatic Stage	Approximate Time Interval After Fertilization	Germ Cells	Follicle
Embryo	25 days	Migration of primordial germ cells and mitotic proliferation of oogonia; first meiotic prophase	Primordial follicl
Embryo	50–60 days	First meiotic prophase	
Fetus	100 days	Oocytes entering diplotene stage	
Puberty	11–14 years	First spontaneous ovulation	Follicular growth from primordial follicle to preovulatory follicle
Maturity		Ovulation; sperm entry; completion of the 2nd meiotic division; zygote	"
Menopause	45–55 years	Oocytes (degenerative)	A few primordial or primary follicles

Formation of Embryo

The sperm and egg must unite by a process called fertilization to produce an embryo. For successful fertilization, both the spermatozoon and the egg must undergo a series of changes. In order to penetrate through the corona radiata cells and the zona pellucida to fertilize the egg, the sperm cells must undergo a process called capacitation. In humans, it takes about seven hours for the sperm to complete capacitation. Morphologically it is impossible to distinguish the capacitated from the noncapacitated spermatozoa. The process of capacitation will allow the sperm cells to undergo the next process called activation. The spermatozoa, once capacitated, are still unable to fertilize the egg until they undergo a final change of activation. Three types of changes occur in spermatozoa during the activation process. First, as a result of membrane fusion the contents of the acrosome (a structure located in the anterior region of the spermatozoan head) become exposed to the exterior. This process is called acrosome reaction. Second, changes occur in the flagellar beat, leading to a single episodic "whiplash" type of beat replacing the undulating wave-like beat. Third, changes occur in the plasma membrane surrounding the post nuclear cap region, thus making it more specialized for membrane fusion.

Acrosome reaction releases acrosomal enzymes which allows the sperm cell to pass through the matrix of corona radiata cells and the zona pellucida. The plasma membrane of the spermatozoon then fuses with the oocyte membrane. As soon as this fusion occurs, two drastic changes take place in the oocyte. First, the second meiotic division, which was arrested at the second meiotic metaphase after ovulation, is completed. Second, cortical granules, which were present at the cortex of the egg, release their contents to the exterior of the egg in a process called cortical granule reaction. These materials then act on the zona pellucida to prevent further penetration by other spermatozoa. This process is called the zona reaction. Within two to three hours after spermatozoon-oocyte fusion or fertilization, the second polar body is expelled. The haploid sets of female and male chromosomes lie in the oocyte cytoplasm. At the completion of the fertilization process, two sets of haploid chromosomes, one from the male and the other from the female, produce two pronuclei each of which is surrounded by a distinct nuclear membrane.

During the first 24-h period of fertilization, there is no sign of mRNA synthesis in the ovum, the one-cell or the early two-cell embryo. The first few days before implantation are extremely hazardous for the embryo because the maternal physiology must be altered to accept the embryo's presence. Failure to do so will cause onset of the menstrual cycle leading to the loss of the embryo. Chemical injury during this stage may lead to abnormal hormonal release or alter the microenvironment of the fallopian tube, causing embryonal death.

Reproductive Hormones

The central nervous system (CNS) plays an integrative role in the reproductive process. The neurons in the hypothalamus of the brain secrete the gonadotropin releasing hormone (GnRH). The anterior pituitary cells have receptors for GnRH. In response to GnRH, the anterior pituitary cells secrete follicle stimulating hormone (FSH) and luteinizing hormone (LH). FSH secretion is necessary for the development of the follicles, while both FSH and LH are needed for their final maturation and a burst of LH is required for ovulation and the initiation of corpus luteum formation. Estrogens, the female sex hormones, are produced by the theca cells and the corpus luteum. Granulosa cells also produce estrogens, which remain in the follicular fluid. Mature corpus luteum secretes progesterone. Progesterone is responsible for changes in the endometrium of the uterus. If the implantation takes place, then the early embryo, is capable of prolonging the life of the corpus luteum. The implanting embryo, which is now called the blastocyst is composed of different types of cells. The syncytiotrophoblast cells of the blastocyst produce a hormone which is called human chorionic gonadotropin (hCG). This hormone is critical for maintaining the progestagenic activity of the corpus luteum. After the formation of the placenta, various hormones are secreted by this organ which become central to the maintenance of pregnancy. Exogenous chemicals altering the placental activity may be damaging or lethal to the fetus.

ROLE OF CYTOGENETICS FOR ASSESSMENT OF REPRODUCTIVE HAZARDS

Cytogenetics is a field of genetics which deals with the morphology and behavior of chromosomes and how this relates to human disease. Birth defects are not uncommon. They occur in approximately 3 to 4% of all pregnancies and can range in degree from mild to severe. A common cause of birth defects is a chromosomal abnormality. A major chromosome abnormality occurs in about one in every 200 live births, and it is estimated that as many as one in every twenty conceptions may contain a chromosome error.

Chromosome errors, either in number or structure, generally occur during, or prior to, conception. Errors may occur in the individual egg or sperm. When a baby is born with obvious physical problems, a blood sample is taken and sent to a cytogenetics laboratory for chromosome studies.

When the blood sample arrives in the laboratory, the white cells are grown in culture and after three days are treated with a chemical which arrests cell division in a stage called metaphase where the chromosomes are most easily studied since the individual features are clearly observable with a microscope at this state. The cells are then burst open, the chromosomes are released on a microscope slide, stained, photographed, enlarged, cut out and then arranged into a specific pattern on a sheet of paper. The arrangement of the chromosomes is called a karyotype and is based on the size and shape of the chromosomes as well as on the specific pattern of bands found on each chromosome. A normal karyotype consists of 46 chromosomes in 23 pairs. The autosomes (nonsex chromosomes) are numbered 1 to 22, and the sex chromosomes consist of two X chromosomes for a female and an X and Y chromosome for a male (Figures 12.3 and 12.4).

The most common chromosome error found in living individuals is where chromosome #21 is present in three copies instead of two (Figure 12.5). This cytogenetic abnormality causes Down syndrome which is one of the most common birth defects and a major cause of mental retardation. Approximately 60-80% of the cases of Down syndrome are due to the egg cell carrying two copies rather than one copy of chromosome #21. Animal studies indicate that such a phenomenon occurs following X-irradiation of female mice. Human epidemiological studies, furthermore, suggest an increased incidence of Down syndrome births following maternal diagnostic irradiation.

In addition to chromosomal aneuploidy, i.e., abnormal chromosome number, radiation in general has been shown to cause chromosome breaks and other aberrations. In itself, chromosome changes are mutations. They serve as indicators that cells have been exposed to agents which can potentially cause genetic damage. These are inherent warnings that carcinogenic or teratogenic processes

KARYOTYPE
Normal Female

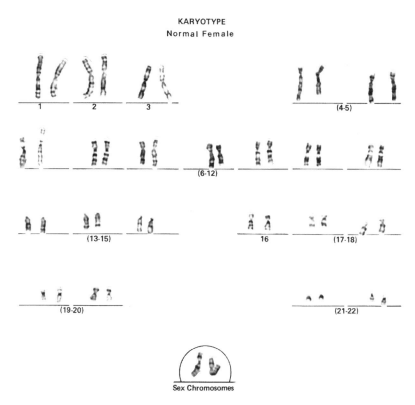

FIGURE 12.3 Normal chromosomes for a female.

may be underway. Approximately half of the leukemias and most of the solid tumors have been shown to have chromosomal abnormalities. One particular abnormality, involving a translocation between chromosome 4 and 11, occurs so consistently in newborns diagnosed with acute lymphocytic leukemia, that researchers now suspect that the neoplastic condition may result from prenatal toxic exposure. Because of the easier accessibility of gametes (spermatozoa) in the male, more compounds have been shown in the past to be toxic to the male gametes (sperm can be scored for shape, mobility, and penetrability) than to the female gametes (eggs). The relative lack of access to ovarian material has made it difficult to examine such issues as female germline exposure to toxins and possible oocyte killing or genetic damage. Recently, however, infertile patients enrolled in "in vitro fertilization/embryo transfer" programs are becoming a source of female gonadal tissue for future studies.

Another source of material, and a noninvasive source, is body fluids such as blood samples from exposed individuals. While investigators cannot confirm the result with absolute certainty, it can be assumed that chromosome changes in the peripheral blood of a female employee may translate to genetic damage

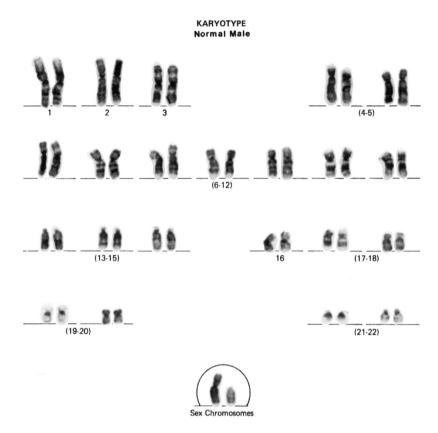

FIGURE 12.4 Normal chromosomes for a male.

to the fetus. Evidence suggests that even peripheral changes should alert researchers to a potential hazard.

Chromosome breakage, besides causing mutations, can also cause death of the cell. While cellular death at the appropriate time is important in normal organogenesis, extensive unscheduled death induced by some exogenous agents could cause abnormal embryonic development. Epidemiological studies clearly indicate that an increased incidence of congenital malformations and spontaneous abortions have been seen in pregnant operating room staff and anesthetists.

The genetic effect of radiation is a serious concern. Two situations need to be considered for radiation effects: (1) the mutagenic effects, resulting in damage to the germ cells before fertilization, and (2) the teratogenic effects, resulting in damage to the developing embryo.

The oocyte is especially radiation sensitive around the time of fertilization. The most usual type of radiation exposure is inadvertent, consists of 1 rad, and is given during the course of investigation for infertility before a pregnancy is recognized. After 1 rad (the upper diagnostic limit) of radiation exposure the added risk to the fetus is approximately 1/1,000 for congenital malformations, mental retardation, and childhood cancer. In contrast, follow-up of

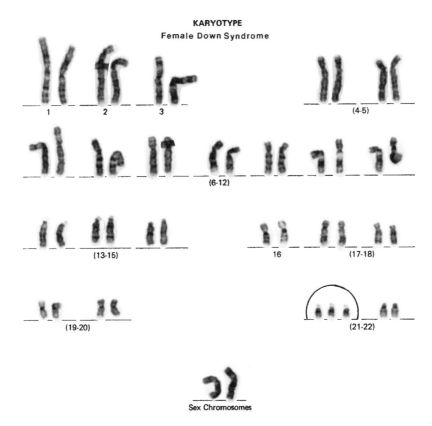

KARYOTYPE
Female Down Syndrome

FIGURE 12.5 Abnormal chromosomes for female down syndrome.

Japanese atomic bomb casualties shows that heavy doses (greater than 20 to 30 rads) lead to a 10% increased risk of mental retardation and microcephaly.

Individuals working with radiation or radioisotopes may justifiably be concerned about the extent of their radiation exposure and the reproductive consequences of this exposure. Currently, there is no well established method for detecting genetic damage occurring in vivo at the level of a single gene or a specific locus, and no tests that can definitively predict that reproductive problems will occur. There are, however, some screening systems available which can be used as indices of exposure and potential danger. One of the methodologies which attempts to monitor occupational reproductive hazards is cytogenetic screening. Based on this information it is clear that cytogenetic analysis has a role to play in assessing occupational reproductive hazards.

EFFECTS OF CHEMICALS ON
FEMALE REPRODUCTIVE PROCESS

About 1000 new chemicals are being introduced each year with inadequate testing for health hazards. The list of chemicals which are known to cause teratogenic effects in humans (Table 12.2) is quite short in comparison to more than 800 chemicals in laboratory animals. This lack of information is due to the inherent difficulty in working with humans. In females, about 45% of the reports on reproductive toxicity deal with embryo and fetotoxicity. Placental toxicity usually causes spontaneous abortions. Occupational exposure to chemicals may alter the integrated function of the hypothalamus-pituitary-ovarian-uterian axis, suggesting some relationship between the chemical exposures and impaired fertility, ovarian toxicity, abnormal hormone production, menstrual disturbance and uterine toxicity. Toxic chemicals can be divided into two categories "direct-acting toxicants" and "indirect-acting toxicants". The direct acting toxicants cause direct damage to the subcellular organelles and macromolecules within the cells (Tables 12.3 and 12.4). While the indirect-acting toxicants alter the metabolic activity of a cell or cause hormonal imbalance (Tables 12.5 and 12.6).

Occupational exposure to sex steroids and contraceptives have been associated with impaired ovarian function and disturbance in the menstrual cycle. Women employed in agricultural work are exposed to pesticides and herbicides. These chemicals contain organohalide and organophosphorus compounds. Both these compounds have been associated with impaired ovarian function, disturbance in menstrual cycle and infertility.

Occupational exposure to carbon disulfide in rayon fiber manufacturing plants is known to cause menstrual disturbance in female workers. Lead, cadmium, mercury, manganese and tin have also been reported to cause abnormal menses and spontaneous abortions. Reports suggest that exposure to gasoline vapors in the industrial setting can alter ovarian functions. Menstrual disturbances were also noted in women exposed to benzene, chlorine, hexachlorocyclohexane and chlorobenzene. Similar observations of altered menses were observed in women exposed to formaldehyde vapors and ethylene oxide. Women employed in plastic industries are exposed to a variety of chemicals including polymerizers, plasticizers and vinyl chloride. These chemicals, too, have been reported to cause menstrual disturbances.

Occupational exposure of chemicals may cause ovarian toxicity by directly destroying oocytes or the follicles or may cause genetic mutations leading to developmental disorders or carcinogenesis (Tables 12.7, 12.8, and 12.9). Depending on the site and mechanism of action of an ovarian toxin, a period of subfertility or infertility may occur. Because of the use of contraceptives it is very difficult to assess the reproductive toxins in human females.

Ideally, evaluation of risks would be greatly facilitated by including multiple end points for analysis rather than just one assay and by engaging in research collaboration across the various disciplines (epidemiologic, biochemical, cytogenetic, and molecular biology).

TERATOGENIC EFFECTS OF INDUSTRIAL CHEMICALS

Since the epidemic of limb defects due to the maternal ingestion of thalidomide, more stringent testing of new agents for teratogenicity has been imposed.

As mentioned before, despite widespread concern, actual evidence for human teratogenic effects of most chemicals is limited (Table 12.2).

TABLE 12.2 Human Teratogens

Known Teratogenic Agents	Possible Teratogens
Ionizing Radiation	Binge drinking
Atomic weapons	Carbamazepine
Radioiodine	Cigarette Smoking
Therapeutic	Disulfiram
Infections	High Vitamin A
Cytomegalovirus (CMV)	Lead
Herpes virus hominis? I and II	Primidone
Parvovirus B-19	Streptomycin
(Erthema infectiosum)	Toluene abuse
Rubella virus	Varicella virus
Syphilis	Zinc deficiency
Toxoplasmosis	**Unlikely Teratogens**
Venezuelan equine	
encephalitis virus	Agent Orange
Metabolic Imbalance	Anesthetics
Alcoholism	Aspartame
Diabetes	Aspirin (but aspirin in the second
Folic acid deficiency	half of pregnancy may increase
Hyperthermia	cerebral hemorrhage during
Phenylketonuria	delivery)
Rheumatic disease and congenital heart block	Bendectin (antinauseants)
Virilizing tumors	Birth control pills
Drug and Environmental Chemicals	Marijuana, lysergic acid
Aminopterin and methylaminopterin	diethylamide (LSD)
Androgenic hormones	Metronidazole
Busulfan	Oral contraceptives
Captropril; Enalapril and renal damage	Rubella vaccine
Chlorobiphenyls	Spermacides
Cocaine	Video display screens
Coumarin anticoagulants	
Cyclophosphamide	
Diethylstilbestrol	
Diphenylhydantoin and trimethadione	
Etretinate	
Lithium	
Methimazole and scalp defects	
Mercury, organic	
Penicillamine	
Tetracyclines	
Thalidomide	
Trimethadione	
13-cis-retinoic acid	
(isotreninoin and Accutane)	
Valproic acid	

TABLE 12.2   Human Teratogens (continued)

Source: Shepard et al., Developmental toxicology: prenatal period, in *Occupational and Environmental Reproductive Hazards: A Guide for Clinicians*, Williams & Wilkins, Philadelphia, PA, 1993. Chapter 4, page 51 with permission of M. Paul (ed.) and Williams & Wilkins, Baltimore, MD.

TABLE 12.3   Direct-Acting Reproductive Toxicants (Chemical Reactivity)

Compound	Effect	Site	Mechanism
Alkylating agents	Altered menses Amenorrhea	Ovary	Follicle toxicity
Lead	Abnormal menses Ovarian atrophy Decreased fertility	Hypothalamus? Pituitary? Ovary?	Decreased FSH Decreased progesterone
Mercury	Abnormal menses	Hypothalamus? Ovary	Follicle toxicity Granulosa cell proliferation
Cadmium	Follicular atresia Persistent diestrus	Ovary Pituitary Hypothalamus	Vascular toxicity Direct toxicity

Source: Plowchalk et al., Female reproductive toxicology, in *Occupational and Environmental Reproductive Hazards: A Guide for Clinicians*, Williams & Wilkins, Philadelphia, PA, 1993. Chapter 2, page 19 with permission of M. Paul (ed.) and Williams & Wilkins, Baltimore, MD.

Table 12.4   Direct-Acting Reproductive Toxicants (Structural Similarity)

Compound	Effect	Site	Mechanism
Oral contraceptive	Altered menses	Hypothalamus Pituitary	Altered FSH, LH release
Azathioprine	Reduced follicle numbers	Ovary Oogenesis	Purine analog
Halogenated hydrocarbons			
Chlordecone	Sterility	Hypothalamus?	Estrogen agonists
DDT[a]	Altered menses	Pituitary?	
2,4-D[b]	Infertility		
Lidane	Amenorrhea		
Toxaphene	Hyperamenorrhea		
Hexachlor			

[a] Dichlorodiphenyltrichloroethane

[b] 2,4-dichlorophenoxyacetic acid

Source: Plowchalk et al., Female reproductive toxicology, in *Occupational and Environmental Reproductive Hazards: A Guide for Clinicians*, Williams & Wilkins, Philadelphia, PA, 1993. Chapter 2, page 19 with permission of M. Paul (ed.) and Williams & Wilkins, Baltimore, MD.

The mechanisms by which chemicals induce teratogenic effects are not well understood. Teratogens are exogenous agents which induce birth defects.

TABLE 12.5　Indirect-Acting Reproductive Toxicants (Metabolic Activation)

Compound	Effect	Site	Mechanism
Cytoxan	Amenorrhea	Ovary	Follicle destruction
	Premature ovarian failure		
Polycyclic aromatic hydrocarbons	Impaired fertility	Ovary	Follicle destruction
		Liver	Enzyme induction
Cigarette smoke	Altered menses	Ovary	Follicle destruction
	Impaired fertility		Blocked ovulation
	Reduced age at menopause		
DDT[a] metabolites	Altered steroid metabolism	Liver	Enzyme induction

[a] Dichlorodiphenyltrichloroethane

Source: Plowchalk et al., Female reproductive toxicology, in *Occupational and Environmental Reproduction Hazards: A Guide for Clinicians*, Williams & Wilkins, Philadelphia, PA, 1993. Chapter 2, page 20 with permission of M. Paul (ed.) and Williams & Wilkins, Baltimore, MD.

TABLE 12.6　Indirect-Acting Reproductive Toxicants (Metabolic Activation)

Compound	Effect	Site	Mechanism
Halogenated hydrocarbons			
DDT[a]	Abnormal menses	Hypothalamus?	FSH
		Pituitary?	LH
PCBs, PBBs[b]	Abnormal menses	Hypothalamus?	FSH
		Pituitary?	LH
		Liver	Enzyme induction
Barbiturates	Increased steroid clearance	Liver	Enzyme induction

[a] Dichlorodiphenyltrichloroethane

[b] Polychlorinated and polybrominated biphenyls

Source: Plowchalk et al., Female reproductive toxicology, in *Occupational and Environmental Reproductive Hazards: A Guide for Clinicians*, Williams & Wilkins, Philadelphia, PA, 1993. Chapter 2, page 20 with permission of M. Paul (ed.) and Williams & Wilkins, Baltimore, MD.

Out of all cases of birth defects, 5 to 10% are caused by teratogenic agents, 20 to 25% are caused by genetic factors and 60 to 65% are of unknown origin.

A Catalogue of Teratogenic Agents, edited by Thomas H. Shepard (1989), has been published in an attempt to recognize potential teratogenic hazards and congenital malformations. Each listing includes a main entry with synonyms. This is followed by a brief account of some of the work including species, dose, gestational age at time of administration, and type of defect. References following each entry were chosen because of their review nature, originality or because they are most current.

It is difficult to extrapolate human risk figures from animal studies just as it is difficult to assess increased incidences of malformations when these malformations are variable or commonly occur in the absence of the agent. Probably

TABLE 12.7 Oocytes as Targets for Chemical Injury

Site of Action	Mechanism of Action (Outcome)
Oocyte maturation	Disrupted communication between oocyte and granulosa cells of the corona radiata (loss of proper biochemical signals for maturation)
	Interference with synthesis and secretion of the zona pellucida proteins (abnormal sperm receptor content → nonviable ovum)
	General cytotoxicity to cellular processes (Oocyte death)
Meiotic maturation	Damage to oocyte DNA
	Disrupted communication with granulosa cells
	Interference with mechanisms that control germinal vesicle breakdown (untimely meiotic divisions

Source: Plowchalk et al., Female reproductive toxicology, in *Occupational and Environmental Reproductive Hazards: A Guide for Clinicians*, Williams & Wilkins, Philadelphia, PA, 1993. Chapter 2, page 23 with permission of M. Paul (ed.) and Williams & Wilkins, Baltimore, MD.

TABLE 12.8 Granulosa Cells as Targets for Chemical Injury

Site of Action	Mechanism of Action (Outcome)
FSH and LH receptors	Decreased receptor population
	Competition for receptor
	Uncoupling of receptor to secondary messenger
	(Decreased estradiol production)
	(Accumulation of androgens → atresia)
	(Inadequate luteinization)
	(Decreased progesterone production)
	(Luteal phase defects)
Steroid production	Altered estrogen production: inhibits or depresses aromatase activity
	(Excessive follicular androgens → atresia)
	Inadequate source of androgens
	(Decreased estrogen → altered follicle growth)
	Altered progesterone production: inhibition of enzymes responsible for biosynthesis of progesterone from cholesterol)
	Inadequate luteinization of granulosa cells
	(Decreased progesterone → inhibition of FSH)
	(Decreased progesterone → luteal phase defect)
Cell proliferation	General cytotoxicity
	Mitotic inhibitors
	Reduced production of growth factors
	(Follicular atresia?)

Source: Plowchalk et al., Female Reproductive Toxicology, in *Occupational and Environmental Reproductive Hazards: A Guide for Clinicians*, Williams & Wilkins, Philadelphia, PA, 1993. Chapter 2, page 22 with permission of M. Paul (ed.) and Williams & Wilkins, Baltimore, MD.

TABLE 12.9 Thecal Cells as Targets for Chemical Injury

Site of Action	Mechanism of Action (Outcome)
LH receptors	Decreased receptor population
	Competition for receptor
	Uncoupling of receptor for secondary messenger
	(Decreased androgen biosynthesis)
	(Insufficient substrate for granulosa cells)
	(Altered follicular growth)
Steroid production	Inhibition of enzymes responsible for biosynthesis of androgens from cholesterol
	(Altered androstenedione and testosterone levels)
	(Insufficient substrate for granulosa cells)
Cell proliferation	Disrupted migration of stroma to form thecal cell layer
	General cytotoxicity
	Mitotic inhibition
	Reduced production of growth factors

Source: Plowchalk et al., Female reproductive toxicology, in *Occupational and Environmental Reproductive Hazards: A Guide for Clinicians*, Williams & Wilkins, Philadelphia, PA, 1993. Chapter 2, page 23 with permission of M. Paul (ed.) and Williams & Wilkins, Baltimore, MD.

the best advice to a pregnant woman is to avoid all drugs that are not strictly essential (including alcohol and cigarettes) and eat a nutritious balanced diet.

End Points Evaluation for Chromosome Damage

Chromosomal changes such as chromosome breakage, aneuploidy, and structural changes may cause no effects or range in effects from death of the cell or spontaneous miscarriage of the fetus, to developmental anomalies or malignancies.

The micronucleus test is a simple *in vivo* and *in vitro* cytogenetic screening method for mutagens. It is usually done in lymphocytes or bone marrow cells. Its most practical application is as an initial screening test in laboratories in which a large number of compounds are to be tested as it is less laborious and time consuming than other established tests for chromosome breakage. In principle, micronuclei originate from chromosome breakage where pieces of chromosomes become surrounded by a membrane. Since micronuclei are easy to score, larger numbers of cells can be counted than in conventional karyotypic analysis.

The sister chromatid exchange technique is another exceedingly sensitive end point test for detecting certain chemical mutagens. Chemicals which may fail to induce chromosome abnormalities, such as breaks and structural rearrangements, can induce sister chromatid exchanges, and chemicals which induce chromosome abnormalities only at high doses have been shown to induce sister chromatid exchanges even at very low concentrations. The sister chromatid exchange technique requires incorporation of BudR, a nucleoside analog for thymidine during two consecutive cell cycles followed by staining and microscopic visualization. Chromosomes which are replicated in the presence of BudR for two cell cycles contain one chromatid which has one strand of substituted DNA (unifilary substituted) and one chromatid with both strands substituted (bifilary substituted). These differences in analog substitution can then be exploited to differentially stain the chromatids. Unifilary substituted chromatids stain dark, whereas bifilary substituted chromatids stain light. Toxic chemicals produce reciprocal alterations or exchanges between sister chromatids which can be scored by counting the number of times light and dark regions exchange places.

REFERENCES

Austin, C. R. and Short, R. V., *Reproduction in Mammals, Book I: Germ Cells and Fertilization*, Cambridge University Press, Cambridge, England, 1982.

Hiller, S. G., *Gonadal Development and Function*, Serono Symposia Publications, Volume 94, Raven Press, New York, 1992.

Knobil, E. and Neill, J. D. (eds), Greenwald, G. S., Marrut, C. L. and Pfaff, D. W. (assoc. ed), *The Physiology of Reproduction*, 2nd edition, Raven Press, New York, 1994.

Paul, M., *Occupational and Environmental Reproductive Hazards: A Guide for Clinicians*, Williams & Wilkins, Philadelphia, 1993.

Sorsa, M. and Vainio, H., *Mutagens in Our Environment*, Alan R. Liss, New York, 1982.

Thompson, M. W., McInnes, R. R. and Willard, H. F. (eds), *Genetics in Medicine*, 5th edition, W. B. Saunders, Philadelphia, 1991.

13

Toxic Responses of the Male Reproductive System

Joana Chakraborty

CONTENTS

INTRODUCTION

The toxic substance can impair the male reproductive system at several sites, including neural, endocrine, gonadal, and accessory glands, and can affect sexual behavior. In the male reproductive system, a number of separate physiological processes must work in a harmonious fashion in order to maintain normal fertility. These physiological processes range from the central and autonomic nervous system to secretory functions of various glands and the endocrine regulatory system. Disturbances at any level may disrupt the normal reproductive capacity of a man. The impact of exogenous chemicals on the testis may cause genetic disorders. It is known that recessive mutations in germ cells (reproductive cells) may be accumulated undetected for generations before being expressed.

Frequently, in toxicological studies, the male reproductive system is not examined in routine autopsy. Even when there are clear indications of the toxic nature of certain chemicals impairing fertility, studies are conducted mainly on females. This is due to an obvious lack of easily measurable reproductive indicators in the male system. For example, in pregnant women, the occurrence of a menstrual cycle may indicate spontaneous abortion or pregnancy wastage. Chemical agents may also appear in breast milk. Therefore, in many cases, breast milk can be regarded as an indicator of reproductive toxicity. Whereas in the male, semen analysis does not provide a definite clue of infertility. There is no good laboratory test for clinical evaluation of human male reproductive function. Testicular biopsy is not performed to investigate the effect of noxious chemicals. As a result, there is no means to study the effect of environmental pollutants on the testis except testing the end points, i.e. sperm production in semen. In the testis, unlike the ovaries, division of cell is a continuous process. Dividing cells are far more sensitive to any toxicant than cells at rest. Therefore, the testis is at greater risk.

This chapter has been divided into three major sections. First, various types of cells and their function in adult human testis will be discussed. Second, the effects of various toxic chemicals on testis will be reviewed. Third, need for further study and validity of current testing procedures in endpoint evaluation will be evaluated.

THE ADULT TESTIS

The adult testes are ovoid organs. Each measures about $4 \times 3.5 \times 3$ cm. They are located at the end of a cord, called the spermatic cord. The testis is freely movable in the scrotal sac. The scrotum is composed of skin, a thin layer of dartos muscle and a layer of connective tissue which includes striated muscle, called cremaster muscle. The testis is surrounded by a thick capsule, the tunica albuginea. Numerous septa are present in the testis, dividing it into smaller compartments or lobules. Each lobule contains two to four highly tortuous seminiferous tubules (Figure 13.1 a, b, and c). The adult testis has dual functions: production of germ cell, i.e., spermatozoa; and secretion of androgens. The germ cells are produced in the seminiferous tubules while the androgen is produced by the cells outside the seminiferous tubules, the Leydig cells.

Seminiferous Tubules, Spermatogenesis and Sertoli Cells

The adult seminiferous tubule is a narrow cylindrical structure with an open central canal, the lumen. The wall of each tubule is lined with layers of specialized epithelial cells. The majority of these cells are germ cells or sex cells at various stages of development and differentiation. A layer of another type of cells, the Sertoli cells, is present between the developing germ cells. These cells rest upon an inner acellular zone, the basal lamina. The other side of the basal lamina is surrounded by a layer of cells called the myoid cells. The myoid cells in turn are surrounded by the adventitial cell layer.

Spermatogenesis starts at puberty with the mitotic division of spermatogonia. The spermatogonia are located next to the basal lamina between the Sertoli cells. However, in the human at the onset of spermatogenesis, all spermatogo-

Sertoli Cell

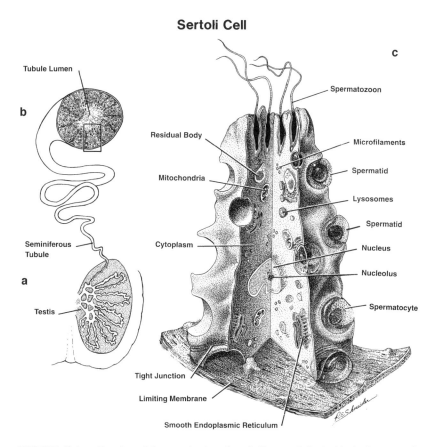

Tubule Lumen

b

Spermatozoon

Residual Body

Microfilaments

Mitochondria

Spermatid

Lysosomes

Spermatid

Seminiferous Tubule

Cytoplasm

Nucleus

Nucleolus

a

Spermatocyte

Testis

Tight Junction

Limiting Membrane

Smooth Endoplasmic Reticulum

c

FIGURE 13.1 a. Drawing of the organization of seminiferous tubules inside the human testis. b. Drawing of an uncoiled seminiferous tubule showing the location of the germ cells at low magnification. c. Enlarged drawing of a portion of the seminiferous tubules. Showing the association of Sertoli cells and germ cells.

nia in the entire cross section of the seminiferous tubule do not divide. Some remain in the resting state, to become active later. Each spermatogonial cell undergoes a limited number of mitotic divisions, producing a clone of daughter cells. There are four types of spermatogonia in the human: spermatogonia type A long, spermatogonia type A dark, spermatogonia type A pale and spermatogonia type B. The spermatogonia may be highly sensitive to some chemical agents and radiation. In rodents, low dosages of irradiation were shown to destroy the spermatogonia. However, a few of them may be highly resistant. These resistant cells were identified as type A stem cells. These same cells may be highly resistant to other noxious agents.

The process of spermatogenesis has been summarized in Figure 13.2. Briefly, spermatogonia Type A produce type B spermatogonia. These cells are particularly vulnerable to extraneous influences, such as antiandrogens and irradiation. Spermatogonia type B divide by mitosis to produce primary spermatocytes. The primary spermatocytes are the ones which will undergo the first meiotic division. The long prophase of the first meiotic division is divided into five stages: leptotene, zygotene, pachytene, diplotene and diakinesis. In the zygotene stage, the appearance of the "synaptonemal complex" is the most significant event. The synaptonemal complexes are associated with the chromosomal pairing before crossing over. Therefore, any external influence that impairs the formation of synaptonemal complexes may interfere with the normal chromosome paring and crossing over. The synaptonemal complexes persist in the pachytene stage. In the diplotene stage the chromosomes begin to move apart from each other, they are connected only at the crossing over sites called chiasmata. In diakinesis, the chromosomes become short and thick. Then at the end of prophase, the chromosomes are arranged at the equator, the characteristic of metaphase. Finally, at the completion of anaphase and telophase each homologous chromosome along with two sister chromatids migrates to one pole. After cytokinesis, two secondary spermatocytes are formed. Within a few hours, the secondary spermatocytes undergo a second meiotic division, resulting in the formation of spermatids. The spermatids are haploid cells. They do not undergo further division. By a series of nuclear and cytoplasmic modifications each spermatid is now transformed to a spermatozoon. This long transformation process is called spermiogenesis. This transformation includes the formation of the acrosome, mitochondrial reorganization, nuclear condensation, tail formation and shedding of the excess cytoplasm with remaining organelles. In humans, the entire process takes about 74 days. However, it has been shown that at least 3 to 4 months are needed for production of spermatozoa in a human testis that has been irradiated, therefore, it can be estimated that the spermatogenesis process takes about 100 days in man. The entire process of spermatogenesis occurs in close association with the Sertoli cells.

The Sertoli cells are relatively few in number. They are tall cells with branch-like projections (Figure 13.1c). Each Sertoli cell extends from the base of the seminiferous tubules to the lumen. The germ cells are present in indentations of the Sertoli cells. In the adult seminiferous tubules, the Sertoli cells do not divide. They play a coordinating role. The Sertoli cells maintain contacts with their neighboring cells via gap junction. Thus they are active in the information transfer system around and along the seminiferous tubules. Sertoli cells maintain the overall architecture of various layers of germ cells; they secrete fluid and some important proteins, including androgen-binding protein; they play an active role in the release of spermatozoa into the lumen of the seminiferous tubule.

The production of spermatozoa and androgens occur in two different compartments of the testis, i.e., spermatozoa are produced within the seminiferous tubules while androgens are synthesized in between the tubules. These two compartments are both morphologically and physiologically separated from each other. Therefore, fluid collected from the lumen of the seminiferous tubules is quite different than the fluid in the intertubular compartment. It is

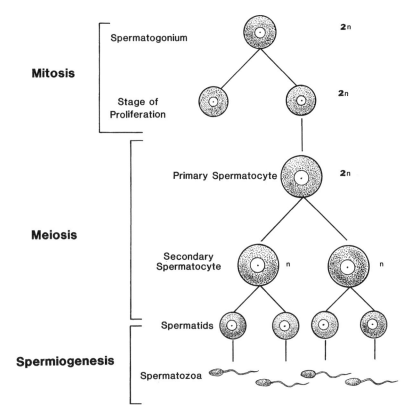

FIGURE 13.2 A schematic drawing of the process of spermatogenesis and spermiogenesis. In this drawing, spermatogonium is shown to contain diploid number of chromosomes (2n). These cells divide mitotically producing diploid (2n) primary spermatocyte cells. The primary spermatocytes undergo meiosis, thus each of them produce 2 haploid (n) secondary spermatocytes. Each secondary spermatocyte produces 2 spermatids (each contain either X or Y chromosomes). The spermatids differentiate by the process of spermiogenesis during which each of them develop to one spermatozoon.

clear that there is a blood-testis barrier which maintains a special microenvironment within the seminiferous tubules. Inside this special microenvironment, the developing germ cells are protected from external noxious substances. This barrier also prevents the germ cell antigen to gain access to the general circulation, thus protecting an individual from becoming autoimmune against his own spermatozoan antigen. The tight junctional complexes between the neighboring Sertoli cells form this blood-testis barrier in humans.

Extratubular Components and Leydig Cells

The coiled loops of the seminiferous tubules are in close association with loose connective tissue components of the extratubular areas. These areas are also highly vascularized. The intertubular areas are occupied by fibroblasts, macrophages, mast cells, small unmyelinated nerve fibers, lymphatic vessels and clusters of interstitial cells, called the Leydig cells (Figure 13.3). The blood vessels in the intertubular space consist of arterioles, capillary network and venules. Groups of capillaries originating from the arterioles form anastomosing networks. The Leydig cells are seen close to the blood vessels. The Leydig cell cytoplasm contains a variable amount of lipid droplets and a large amount of smooth endoplasmic reticulum. Human leydig cells contain some crystalline inclusions, called the Reinke's crystals. The functional significance of these crystals is not known, however, they appear in the Leydig cells with lowered steroid production.

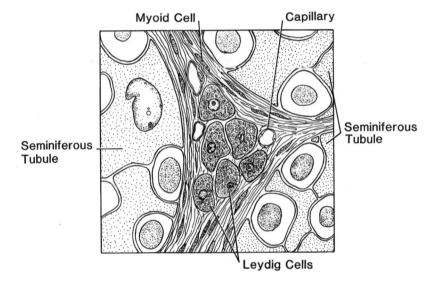

FIGURE 13.3 Extratubular components of the testis showing the location of Leydig cells, capillaries and myoid cells in relationship to the seminiferous tubules.

EFFECTS OF TOXIC CHEMICALS ON TESTIS

The testis may be damaged by a wide range of chemicals. There are two major problems with the study of reproductive toxicity of an agent:

1. Whether there is a threshold level for each type of agent;
2. Whether extrapolation of laboratory data from animal experimentation should be applied for evaluation of risk factors in the human.

For example, thalidomide, a human teratogen is only a weak teratogen in animals. Similarly, aspirin a strong rodent teratogen has no such effect on human.

The gonads play a crucial role in developing the reproductive capacity of a man. Adverse interactions of testicular cells with exogenous agents may cause germ cell mutation, cell death, gonadal tumor, impaired hormone production and loss of fertility. A number of chemicals are known to cause damage or have been suspected to cause damage to the male reproductive process (Table 13.1), however, this is a very short list.

In 1977, the discovery of infertility and sterility among male employees exposed to dibromochloropropane (DBCP) is one example of the dangerous consequences of the lack of systematic investigation. DBCP is a liquid nematocidal agent, that was widely used to save perennial crops without damaging the plant. In the United States, citrus, grapes, peaches, pineapple, soybeans and tomatoes used to be treated with DBCP. Since 1979, DBCP use has been stopped in the United States. The toxic effect of DBCP was not discovered by the scientists and physicians. The workers themselves noted the paucity of children of the men exposed to DBCP. After this problem was reported, a thorough investigation was conducted. In seven separate studies, it was shown that occupational exposure to DBCP caused testicular damage. Testicular biopsies of men exposed to DBCP showed spermatogenic damage. It was suggested that DBCP directly damaged the spermatogonia. In some severely affected men, many seminiferous tubules had no germ cells. Those tubules had only Sertoli cells. Some men became permanently azoospermic.

Dimethylformamide (DMF) is a chemical used in the production of paint, artificial fibers and drugs. In leather dyes and pesticides, it is used as a solvent. Over 100,000 workers in the United States are exposed to this agent. Men employed in the leather tanning industry and some aircraft repairmen has been reported to have testicular cancer. The testicular cancer is not rare among young adult men in the United States. In fact, the rate of this type of cancer appears to be increasing in the past several decades. Further work is needed to verify these initial findings of the association of DMF with testicular cancer.

Another example is the 2,3,6,8-tetrachlorodibenzo-p-dioxin or TCDD (Dioxin). Exposure to TCDD present in Agent Orange, by the Vietnam veterans remained unresolved. Although there were reports of increasing congenital abnormalities among children fathered by the exposed veterans compared to the children fathered by unexposed individuals, according to U.S. government reports these data are not conclusive. Infertility data are also inconclusive.

A study was conducted on the relationship between paternal occupational exposures to six commonly used organic solvents, including styrene, toluene, xylene, tretrachloroethylene, trichloroethylene and 1,1,1-trichloroethane, on spontaneous abortions and congenital malformations. Some indications of increased risk of spontaneous abortion was noted as a result of preconceptional paternal exposure to these solvents. However, due to the lack of a significant number of cases, these results were not conclusive.

A study on the risk of birth defects among offspring of firemen showed that firemen had an increased risk of having a child with ventricular or arterial septal defects. Although this study was carried out on a large number of cases, some limitations in interpretation of results should not be ignored. For example, factors such as maternal drug use, infection or occupation would increase the risk of congenital heart defects. Therefore, further information is needed to obtain a definite correlation between paternal employment as a firefighter

TABLE 13.1   Agents and Substances Reviewed for Reproductive Health Effects by Office of Technology Assessment

Metals	Physical Factors
Lead	Ionizing radiation:
Boron	X-rays
Manganese	Gamma rays
Mercury	Nonionizing radiation:
Cadmium	Ultraviolet radiation
Arsenic	Visible light
Antimony	Infrared radiation
	Radiofrequency/microwave
Chemicals	Laser
Agricultural chemicals:	Ultrasound
Carbaryl	Video display terminals
Dibromochloropropane (DBCP)	Magnetic field
DDT	Hyperbaric/hypobaric environments
Kepone (Chlordecone)	Cold environments
2,4,5-T Dioxin (TCDD) and	Noise
Agent Orange	Vibration
2,4-D	
Polyhalogenated biphenyls:	**Biological Agents**
Polybrominated biphenyls (PBB)	Rubella
Polychlorinated biphenyls (PBC)	Cytomegalovirus
Organic Solvents:	Hepatitis
Carbon disulfide	Other infectious agents
Styrene	Recombinant DNA
Benzene	
Carbon tetrachloride	
Trichlorethylene	
Anesthetic Agents:	
Epichlorohydrin	
Ethylene oxide (EtO)	
Formaldehyde	
Rubber manufacturing:	
1,3-Butadiene	
Chloroprene	
Ethylene thiourea	
Vinyl halides:	
Vinyl chloride	
Undefined industrial exposures:	
Laboratory work	
Oil, chemical and atomic work	
Pump and paper work	
Textile work	
Agriculture work	

Source: U.S. Congress, Office of Technology Assessment, Reproductive Health Hazards in the Workplace, Page 7, OTA-BA-266 (Springfield, VA: National Technical Information Service, December 1985).

and a child's risk of having a septal defect. Paternal occupation as a welder, machinist or metal worker was reported to increase the risk of having retinoblastoma in their children. Association of the father's occupation as an electrician, asbestos worker, electronics worker or printer and an increased risk of offspring's neuroblastoma was also reported.

The adverse effect of lead on male reproduction has long been known. In fact, it dates back to ancient Rome, where lead in drinking water vessels of the upper class Romans were associated with the declining fertility. It was shown that lead had a direct effect on spermatogenesis, causing abnormal or low sperm production thus causing male infertility. Lead also was reported to cause chromosomal abnormalities. Cadmium has also been known specifically for many years to cause testicular damage. Although the effects of cadmium on the human testis have been studied for the last twenty-five years by a large number of investigators, the results are confusing. Therefore, it appears that male reproductive toxins are difficult to identify and difficult to study. Table 13.2 is a partial list of some chemicals and their effects on male reproduction.

NEED FOR FURTHER STUDY AND VALIDITY OF THE CURRENT TEST PROCEDURE

The process of toxicity involves the absorption and distribution of the toxic agent. This event is followed by the toxification and detoxification through the metabolic processes which lead to the excretion of the toxin followed by the repair of damaged tissues. Therefore, in order to assess the toxicity of a chemical on the gonads, a number of factors must be considered. These include: permeability of the agent in question, existence of a biological barrier such as the blood-testis barrier, and whether it affects the germ cell or somatic cell. If it is the germ cells, then the extent of genetic damage, site of biotransformation of the chemical agent, the quality and quantity of the metabolites leading to the increase or decrease of the toxicity of the agent must be determined. The major problem of assessing testicular risk due to occupational exposure of noxious substances stems from the lack of basic understanding regarding the normalcy of the human testis. It is extremely difficult to evaluate subtle damage in the human testis. While animal testicular specimens are being studied thoroughly after experimental exposure to noxious agents, only occasional reports on human testes are found in the literature based on crude evaluations of histologic preparations, sperm output, fertility and measurement of hormone levels in blood. No systematic training program is available in Andrology and only a few investigators are involved in the research of human testicular functions. When human testicular biopsies are available for evaluation, in most cases, the tissue preparation may be done by untrained personnel, leading to the erroneous interpretation of histopathology, in most cases. In man, the lack of basic knowledge on testicular toxins is mainly due to the problem in moral, ethical and legal issues and the lack of good experimental design and adequate procedures.

TABLE 13.2 A Partial List of Chemical Agents Which Affect Male Reproduction

Agents	Degree of Effect (Definite, Inconclusive or No Adverse)	Type of Effect on Male Reproduction	Various Uses of The Chemical Agents
Anesthetic gases	Inconclusive	Infertility	Operating Room
Arsenic	Inconclusive	DNA damage	Smelting industry
Benzene	Inconclusive	Reproductive effects	Solvent, crude oil additive
Boron	Inconclusive	Infertility	Boric acid factory
Cadmium	Inconclusive	Direct effect on spermatogenesis	Alloy, alkaline storage batteries, pigment
Carbaryl	Inconclusive	Oligospermia, abnormal sperm morphology	Insecticides
Carbon disulfide	Definite adverse effect	Spermatogenic damage	Rayon manufacturing
Chlordecone (Kepone)	Inconclusive	Spermatozoan abnormality	Insecticide
Chloroprene	Inconclusive	Affects spermatogenesis	Synthetic rubber
Dibromochloropropane (DBCP)	Define adverse effect	Azoospermia, oligospermia, increased serum FSH, LH	Soil fumigant to control nematodes paint, artifical fibers, drugs
Dimethylformamide (DMF)	Inconclusive	Testicular cancer	Solvent for leather
Solvents (hydrocarbons and glyco ethers)	Inconclusive	Spermatogenic damage	Petroleum and
Dinitrotoluene (DNT) and Diamine (TDA)	Inconclusive	Reduction in sperm count	Chemical plant
Ethylene dibromide	Inconclusive	Infertility ?	Fumigant compound in leaded gasoline
Lead	Definite adverse effect	Spermatogenic damage	Battery stained glass, paint, gasoline
Glycerine	No adverse effects		
Manganese	Inconclusive	Psychomotor and neurological distur-bances	Steel manufacturing

Methylmercury	Inconclusive	Decreased sperm count	Mine industry
P-tertiary butyl benzoic acid (P-TBBA)	No adverse effects		Cutting oil and paint
Pesticides	Inconclusive	y-chromosome breakage, decreased spermatogenesis	Pesticide
PBB	No adverse effects		Fire retardants in plastics
PCB	Inconclusive		Paints, plastics
PCP	Inconclusive	?	Wood preservation
TCDD (dioxin)	Inconclusive	Decreased sperm count	Herbicide manufacturing
Vinyl chloride	Inconclusive	Chromosomal abnormalities	Polyvinylchloride manufacturing

Quantitative procedures along with the application of a modern technical approach may provide better understanding. Measurement of hormone levels including LH, FSH, testosterone and prolactin are currently being used by the National Institute for Occupational Safety and Health (NIOSH) for evaluating potential reproductive problems in males, exposed to hazardous chemicals. However, due to the fluctuating nature of testosterone and pulsatile occurrence of LH in circulation one must be careful in sampling. Better testing procedures should be developed for adequate interpretations of the effects of toxins on the human testis.

END POINTS EVALUATION

It is difficult to determine the toxic end points of human male reproductive functions. The ultimate end point evaluation is based on the question whether an individual can father a child. Therefore, so-called established end point data including sperm density, motility, morphology, and serum hormonal level assays may become invalid if a normal pregnancy occurs. However critical semen analyses following strict criteria (Table 13.3) together with hormonal measurements may provide good end point data.

TABLE 13.3   Semen Profile for Assessing Reproductive Toxicant Effects

Sperm concentration
Sperm viability
  Vital stain
  Hypo-osmotic swelling
Sperm Motility
  Percent motile
  Curvilinear velocity
  Straight-line velocity
  Linearity
  Lateral head amplitude
  Beat cross frequency
Sperm size and shape
  Morphology
  Morphometry
Semen parameters
  pH
  Volume
  Marker chemicals from glands
  Toxicant or metabolite concentrations

Source: Schrader and Kesner: Male reproductive toxicology, in *Occupational and Environmental Reproductive Hazards: A Guide for Clinicians*, Williams & Wilkins, Philadelphia, PA, 1993. Chapter 1, page 11 with permission of M. Paul (ed.) and Williams & Wilkins, Baltimore, MD.

Occupational exposure to chemicals is becoming a special challenge to the reproductive biologists and toxicologists. Figure 13.4 summarizes sites of action of some toxic chemicals. Human reproduction is a complex biological process. The reproductive system contains numerous susceptible targets for injury. The injury may be genetic, hormonal or cellular. The genetic toxicity is especially important because it affects the future generations. Many industrial chemicals may not be hazardous individually, but the effects of these chemicals must be understood. Critical evaluations of effects of chemicals on the reproductive system are of great importance for the safety of the current population as well as the future generation. Multidisciplinary approach and the involvement of both scientific and industrial community are crucial to acquire proper knowledge. More epidemiological studies and collection of data on workers in various industrial settings are important. Potential animal models for prediction of human reproductive hazards should be developed. Only through proper knowledge, the means of safe handling of chemicals can be established which is urgently needed.

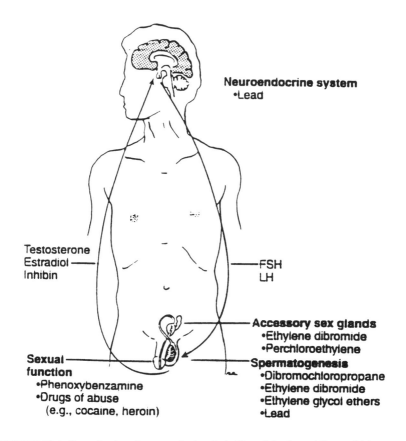

FIGURE 13.4 Sites of action of some toxic chemicals. From Schrader and Kesner: Male reproductive toxicology, in *Occupational and Environmental Reproductive Hazards: A Guide for Clinicians*, Williams & Wilkins, Philadelphia, PA, 1993. Chapter 1, page 4 with permission of M. Paul (ed.) and Williams & Wilkins, Baltimore, MD.

REFERENCES

Barlow, S. M. and Sullivan, F. M., *Reproductive Hazards of Industrial Chemicals*, Academic Press, London, 1982.

Dixon, R., *Reproductive Toxicology*, Raven Press, New York, 1985.

Hiller, S. G., *Gonadal Development and Function*, Serono Symposia Publications, Volume 94, Raven Press, New York, 1992.

Knobil, E. and Neill, J. D. (eds), Greenwald, G. S., Marrut, C. L. and Pfaff, D. W. (assoc. eds), *The Physiology of Reproduction*, 2nd edition, Raven Press, New York, 1994.

Paul, M., *Occupational and Environmental Reproductive Hazards: A Guide for Clinicians*, Williams & Wilkins, Philadelphia, 1993.

Russell, L. D. and Griswold, H. D. (eds), *The Seretoli Cell*, Cache River Press, Clearwater, FL, 1993.

14

Toxic Responses
of the Cardiovascular System

Peter J. Goldblatt

CONTENTS

INTRODUCTION

Despite many advances, cardiovascular disease remains the leading cause of mortality and morbidity in the United States (Table 14.1). While there are hopeful signs that increasing awareness of the role of diet, exercise, and good control of disorders such as essential hypertension and diabetes mellitus are having an impact on the death rate from coronary arteriosclerosis and myocardial infarction, heart disease still ranks as the number one killer in the United States. Contrasted with the increasing awareness of the public of the impact which lifestyle may have in the pathogenesis of cardiovascular diseases, is the relative paucity of information in the medical literature which pertains to the role of the environment and occupation on the development of diseases of the heart and major blood vessels. Despite the growing public and medical awareness of the role of stress in our daily lives, and concerns over the release

1-56670-239-9/97/$0.00+$.50
© 1997 by CRC Press, Inc.

of toxic substances into our environment, relatively little substantiation of the effect of occupations on the cardiovascular system is available so far. Nonetheless, we will attempt to review the available literature, and to relate some of the known pathophysiologic influences to specific occupations. It must be pointed out that some of these relationships are at present only speculative, but it is hoped that these comments will provoke some critical thinking which will lead to the documentation that is required.

TABLE 14.1 Mortality for Leading Causes of Death (United States, 1985)

Rank	Cause of Death	Number of Deaths	Death Rate/100,000 Population	Percent of Total Deaths
	All Causes	2,086,440	739.0	100.0
1.	Heart diseases	771,113	261.4	37.0
2.	Cancer	461,563	170.5	22.1
3.	Cerebrovascular diseases	153,050	51.0	7.3
4.	Accidents	93,457	36.0	4.5
5.	Chronic obstructive lung disease	71,047	25.0	3.4
6.	Pneumonia and Influenza	67,615	22.0	3.2
7.	Diabetes Mellitus	36,969	13.1	1.8
8.	Suicide	29,453	11.2	1.4
9.	Cirrhosis of liver	26,767	10.6	1.3
10.	Arteriosclerosis	23,926	7.6	1.1
11.	Nephritis	21,349	7.3	1.0
12.	Homicide	19,893	7.5	1.0
13.	Diseases of infancy	19,246	8.8	0.9
14.	Septicemia and Pyemia	17,182	6.0	0.8
15.	Aortic aneurysm	15,112	5.3	0.7
	Other and ill-defined	258,698	96.1	12.5

Source: Vital Statistics of the United States, 1985.
From *CA-A Cancer Journal for Clinicians*, Vol. 38, No. 1, January-February.

We shall first turn our attention to the heart, with its major arterial supply (the coronary arteries) and then discuss the aorta and its major musculoelastic branches. We will not dwell on small vessels since the majority of the effects of diseases of these vessels are manifested in the specific organ system that is affected. However, a few examples will be commented upon briefly in the last section.

ENVIRONMENTALLY OR OCCUPATIONALLY INDUCED DISEASES OF THE HEART AND ITS MAJOR BLOOD VESSELS

Pathologic changes of the heart are usually described in relationship to the principal layer of the heart affected (endocardium, myocardium, epicardium). In addition, the coronary arterial supply of the heart is clearly responsible for its most frequent disease in this country, myocardial infarction. In general, environmentally or occupationally related diseases which affect the heart are most likely to arise either in the coronary vessels or in the myocardium, although this is frequently judged on clinical criteria without pathologic confirmation. The major disease of the coronary arteries, atherosclerosis, will be discussed in relationship to environmental influences in the sections dealing with the aorta. Nonetheless, a few remarks with reference to ischemic myocardial damage, as a result of coronary artery disease, and its possible relation to occupation will be commented upon below under the paragraph on myocardium.

Endocardium

The endocardium consists of a single layer of endothelial cells and an under-lying scanty loose connective tissue. The principal diseases affecting the endocardium are more likely to affect the extensions of the endocardium over the atrio-ventricular (mitral and tricuspid) and the pulmonic and aortic valves. Isolated endocarditis, not related to valvulitis, is exceedingly uncommon, and has not been related to occupational exposures. The principal valvular heart disease, rheumatic endocarditis, is declining in incidence in the United States, due to the effective treatment of its etiologic agent, the beta hemolytic Strep-tococcus. Epidemiologic studies have suggested that rheumatic fever is gen-erally a disease of the lower socioeconomic groups, and is related to crowding in an urban industrial environment. Since the disease is principally a disease of children, and is rare as a primary disease of adults, it appears unlikely that occupational influences are of major significance in this disorder. While any traumatic injury which may lead to sepsis could predispose to the development of bacterial endocarditis, especially in individuals with previous rheumatic endocarditis or congenital valvular deformity, occupations in which secondary infection of ulcerated or incised areas of the skin are common might predispose to this condition. Bacterial endocarditis which is not uncommon in intravenous drug users could be classed as an avocational hazard. This may, however, be extending the relationship of occupation to a specific disease beyond the reasonable definitions of occupational relationship. Evidence of industrially related or occupationally related exposures to specific agents which affect either the endocardium or the valves, have not been made. Traumatic rupture of the chordae tendineae of the atrio-ventricular valves is possible as a con-sequence of crushing injury to the chest.

Myocardium

Among the most frequent effects on the myocardium is ischemic necrosis, secondary to coronary vascular occlusion. The relationship between occupation and coronary vascular disease will be commented upon further below, but every-one who watches television is probably convinced that occupational stresses and emotional turmoil are the principal cause of acute myocardial infarction!

In recent years, a number of primary effects on the myocardium have been classified, principally on a clinical basis. These include congestive, hypertrophic, restrictive and obliterative cardiomyopathy. The most common form is classified as "congestive" because of its principal manifestation which is congestive heart failure. Useful as this pathophysiological classification is, the pathologic lesions which lead to enlargement and cardiac failure are many, and may produce a variable clinical pathologic picture. Then too, histopathologic changes in the heart muscle often lag far behind the clinical symptoms of the patient, and may be entirely lacking in the first few hours following even a fatal ischemic episode. On the other hand, some toxic and metabolic injuries produce myofibrillar degeneration which actually appears more rapidly than the manifestations of ischemic injury. Thus, a selective form of myocardial cell necrosis may appear even minutes after the initial injury (contraction band necrosis).

In addition to the myocardial degeneration, the interstitial fibrous connective stroma may show changes which indicate the nature of the pathologic stimulus. Inflammatory cells usually enter the myocardium from the small penetrating intramuscular blood vessels, especially the capillaries, and the characteristic acute inflammatory cells, or lymphocytes and plasma cells may be important clues to the nature of the pathologic process. Edema and hemorrhage may also be seen especially between myocardial fibers and in the perivascular bundles. Fatty change in cardiac myocytes and in the stroma is a consequence of toxic exposure to agents such as ether, alcohol and phosphorus. A cell which is probably a macrophage which has been classically associated with rheumatic fever, the Anitchkov myocyte, may also be seen in other forms of myocardial injury although it is less frequently observed, and the characteristic giant cells are much more frequent in association with rheumatic fever.

Table 14.2 indicates a number of agents which have been associated with myocardial injury. These range from drugs such as the anthracyclines (adriamycin) to trichloroethylene and the fluoroalkane gases used as aerosol propellants. Heavy metals such as lead, and cobalt have also been documented to have effects on the myocardium and these clearly can result from occupational exposure. Cobalt is particularly interesting, in that it has been suggested to have resulted from excessive beer intake when cobalt was used as a "foaming" agent. Another heavy metal which may cause myocardial injury as a result of occupational exposure is mercury. While the effects of mercury on the central nervous system are better known, such as those characterizing the "Mad Hatter" in Alice in Wonderland, mercury has been demonstrated to have deleterious effects on cardiac function and to cause myofibrillar degeneration at least in Rhesus monkeys in experimental exposure. Methyl mercury was used in these experiments and there is evidence that metallic mercury is converted to methyl mercury in the environment. A potentially lethal occupational exposure which has frequently been observed, is to carbon monoxide with its well known binding to hemoglobin and consequent ischemic injury to the myocardium. Firemen, miners and tunnel workers who are exposed to automobile exhaust are particularly at risk. A number of other compounds which may have harmful effects on the myocardium resulting from an occupational exposure are listed in Table 14.2.

TABLE 14.2 Occupational and Environmental Cardiotoxic Agents

Agent	Principal Effect
Inorganic	
Antimony	Sudden death, EKG changes
Arsenic	? Coronary artery disease, multiple
Cadmium	? Hypertension
Cobalt	Cardiomyopathy, pericardial effusion
Lead	? Hypertension
Mercury (also methyl mercury)	Myofibrillar degeneration
Phosphorus	Myocardial fatty change
Organic	
Alcohol	Cardiomyopathy, fatty change
Chloroform	Myocardial fatty change
Carbon disulfide	Coronary artery disease
Carbon monoxide	Myocardial ischemia
Ether	Myocardial fatty change
Halogenated hydrocarbons	
Fluoralkanes	Cardiac arrthymia
Trichloroethylene	Cardiac arrthymias
Nitroglycerine	Coronary arterial spasm (withdrawal)
Others	
Noise	Hypertension
Radiation	
X-ray	Myocardial fibrosis
Microwave	Hypertension, bradycardia
Heat Stress	Hypotension
Iatrogenic	
Anthrycyclines	Cardiomyopathy

Epicardium

Epicarditis is less frequently a cause of significant cardiac malfunction, than diseases of the myocardium. However, obliteration of the epicardial space with a calcific or constrictive kind of pericarditis may result in diminished cardiac filling and consequentially in reduced cardiac output. The close approximation of the pericardial sac to the pleura, allows access of inhaled materials, presumably in part through adjacent lymphatics in the pleural space, and conceivably can result in pericarditis by a mechanism similar to that of infectious pericarditis. Thus, anthracotic pigments and inorganic compounds may occasionally be found to produce areas of inflammation and fibrosis in the pericardial and epicardial tissues. Clearcut evidence of myocardial dysfunction from occupational exposures to compounds such as the silicates and asbestos have been infrequently reported, but it appears possible that massive exposure to these agents may result in pericardial inflammation as well as compromised pulmonary function. Parenthetically, the introduction of asbestos fibers purposefully into the pericardial sac to induce vascular proliferation (granulation tissue) gained some popularity in the early 1950s as a possible treatment for coronary vascular disease. There have been no reports of primary mesotheliomas arising in the pericardial sac as a result of these "Beck" operations. An alleged occupational lesion, "soldier's plaque," probably is the result of healed epicarditis, rather than trauma from backpack straps as was suggested previously.

Chambers

As indicated above, failure of the myocardium can lead to congestive cardiomyopathy which is usually accompanied by ventricular dilatation. Frequently, both chambers are enlarged (biventricular dilatation) and the wall may be thinned or thickened (hypertrophic). While the atria may also dilate, this is unusual in the absence of valvular stenosis or insufficiency. Right ventricular dilatation, usually accompanied by thickening of the myocardium, frequently accompanies chronic pulmonary disease (cor pulmonale) and thus is a complication of occupational lung disease (e.g., silicosis) of longstanding.

Coronary Arteries

The principal disease of the coronary arteries is atherosclerosis. The relationship of atherosclerosis to occupation has not been as clearly drawn as to other risk factors (Table 14.3). Principal among these has been cigarette smoking, high fat diet, and hypertension which will be commented upon further below. As mentioned previously, however, emotional stress appears likely to be a factor in inducing angina pectoris at least in individuals with significant coronary artery narrowing from atherosclerosis. The so-called Type A personality is pictured as a tense, cigarette smoking, deadline-driven individual who rises to the top of the executive ladder over the dead bodies of his colleagues, only to fall victim to a sudden heart attack. While most people think of occupational disease in terms of occupational exposures or physical stresses, indeed the more sedentary and sheltered life of the executive may be the predisposing factor to atherogenesis. An extensive study of longshoreman indicated that heavy manual labor had a protective effect, and even protected against sudden death (Figure 14.1). The role of job-related stress in hypertension remains somewhat speculative, but will be commented further. Less speculative are the studies which show that viscous rayon workers exposed to carbon disulfide for five or more years had an excess of coronary artery deaths more than twice the rate of age-matched nonexposed male controls. Another agent which has been associated with sudden cardiac death possibly from coronary artery narrowing is nitroglycerine. Workers in the explosive manufacturing industry have been noted by several investigators to have sudden death, presumably from coronary artery occlusion, a few days after their last exposure to glycerol nitric esters on their job. The supposition is that this is a vasospasm which occurs as a rebound phenomenon. The use of this compound by many angina-prone individuals to ward off attacks of chest pain is, of course, well known.

It is curious that death as a rebound phenomenon after multiple nitroglycerine tablets is not well documented. This may be due to the fact that the individuals who are affected are anticipated to have sudden coronary occlusions, and no one has paid attention to this phenomenon.

TABLE 14.3 Risk Factors for Coronary Artery Disease

Primary
  Elevated serum cholesterol
  Elevated blood pressure
  Cigarette smoking
Secondary
  Elevated serum or plasma triglycerides (ratio of HDL/LDL)
  Diabetes Mellitus
  Obesity
  Physical inactivity
  Stress: Type A personality

This brief summary of some the occupational conditions which affect the heart directly, will serve to indicate that greater study of the effects of the workplace on the incidence and severity of heart disease are clearly needed.

DISEASES OF THE MAJOR BLOOD VESSELS

While the studies reported below will concentrate on the principal disease of the aorta, namely atherosclerosis, attention will be focused on the condition of essential hypertension, which affects not only the aorta and its major branches, but the small vessels in organs such as the kidney, the brain, and the heart itself. While occupational exposure as a major risk factor for the development of either atherosclerosis or hypertension is somewhat speculative, it is clear that this is a fertile field for some good epidemiologic studies because they are major killers of the Western world.

Table 14.3 depicts the major risk factors in the pathogenesis of atherosclerosis. While the term, arteriosclerosis, is frequently used interchangeably with atherosclerosis, from a pathologist's standpoint there are differences between these two conditions. Atherosclerosis is a disease of the *intima* of large elastic and musculoelastic arteries, characteristically affecting the aorta and its major branches (the coronary arteries, the carotid arteries, the major extrarenal vessels and the vessels of the lower limb principally). It is characterized by fatty deposits which become organized by fibrous connective tissue and may ulcerate or become calcified (complicated lesions). Arteriosclerosis, on the other hand, is a disease of medium sized and small muscular arteries, and arterioles. It is characterized by a thickening of the medial coat and deposition of eosinophilic glass-like material which is generically called "hyaline". The latter condition is much more widespread in individuals with hypertension, and it

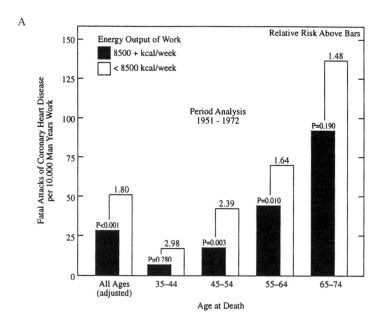

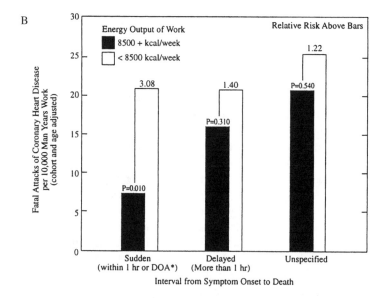

FIGURE 14.1   Relationship of manual labor to fatal heart attack. A. Fatal attacks of coronary heart disease per 10,000 man-years of work among San Francisco longshoremen. 1951–1972. By work energy output and age at death. B. Fatal attacks of coronary heart disease per 10,000 man-years of work among San Francisco longshore-men. 1951–1972. By work energy output and interval from symptom onset to death. (DOA = dead on arrival) From Paffenbarger, R.S. in *Coronary Heart Disease, Prevention and Treatment*, Connor W.E. and Bristow, J.D., Eds., J.B. Lippincott, Philadelphia, 1985, p. 137. With permission from Lippincott-Raven Publishers.*

is classically found in the intrarenal blood vessels especially in association with so-called essential hypertension. A great deal of investigation into the pathogenesis of atherosclerosis has been accomplished, and the role of cholesterol in the diet, high blood pressure, and smoking are clearly defined. Diabetes mellitus is another predisposing factor, but may also be involved in what has been characterized as arteriosclerosis. The effect of the personality, at least on the coronary arteries, has been commented upon above, but there is no question that atherosclerosis appears to be a disease of the industrialized countries, is related to the individual's pattern of exercise, and other patterns of daily living. It is, therefore, clearly possible that certain occupations may predispose individuals to the development of atherosclerosis. Nonetheless, it has been difficult to sort out the effect of a specific occupation from the multiple effects of the other factors which have been fairly convincingly documented to play a role.

Another intriguing area for further investigation, is the relationship of occupation to the development of so-called essential hypertension. High blood pressure is variously defined (Table 14.4), but a number of recent studies have documented that individuals with only mildly increased blood pressure (diastolic above 90 and systolic above 140 mm Hg) are at risk for coronary artery disease and stroke. This has led to recent recommendations to treat even what is called borderline hypertension (Table 14.4) with the contention that this lowers the probability of serious complications. A recent study involving a large number of individuals, indicates that whereas there may have been an effect of antihypertensive therapy on the incidence of stroke, the protective effect of lowering the blood pressure on the coronary arteries was clearly less well documented. Hypertension has been characterized as the silent killer, and close to 40 million individuals in the United States may be affected. Of these, nearly 90% have the form of high blood pressure known as essential hypertension, for which the cause is not known. The role of diet, including such things as alcohol consumption, salt intake and obesity have been widely speculated upon. There appears to be a familial or genetic tendency to develop high blood pressure, and the race and sex of the individual appears to be important at least as predictors of complications. The effect of the sympathetic nervous system to regulate the blood pressure leads inevitably to questions about whether environmental stress and/or psychoneurogenic mechanisms play a major role in a disease for which there is no uniformly accepted pathophysiologic mechanism. It is therefore easy to speculate in the face of vast amounts of ignorance that specific occupations may play a role in either initiating or aggravating abnormal mechanisms of blood pressure regulation. While this has been speculated upon by a number of individuals, documentation is sparse. Nonetheless, it behooves every physician who is interested in the effects of occupation on the cardiovascular system to follow the blood pressure of patients who present particularly with elevated diastolic pressure, and to obtain a good occupational history.

TABLE 14.4 Classification of Raised Blood Pressure

	Blood Pressure	
Classification	Diastolic (mm Hg)	Systolic (mm Hg)
Normal	<85*	<140
High normal	85–89	140
Mild hypertension	90–104	140–180
Moderate hypertension	105–114	>140
Severe hypertension	>115	>140
Isolated systolic hypertension		
Borderline	<90	<140–159
Elevated	<90	160

* WHO accepts 90 as upper limit of normal.

PERIPHERAL VASCULAR DISEASE

As indicated above, small vessel disease as an occupational or environmental hazard is usually discussed in relationship to the specific organ affected. Thus, nephrosclerosis which is present in at least 90% of people with essential hypertension is better discussed with renal disease. One form of peripheral vascular disease that appears to be clearly related to occupation is so-called "white finger" disease accompanied by Raynaud's phenomenon in individuals who have significant vibrational trauma to the hands such as jackhammer users.

REFERENCES

Connor, W. E. and Bristow, J. D. (eds), *Coronary Heart Disease: Prevention, Complications and Treatment*, J. B. Lippincott, Philadelphia, 1985.

Harlan, W. R., Sharrett, A. R., Turino, G. M., Bornani, N. O., and Resnekov, L., Impact of environment on cardiovascular disease. Report of the american heart association task force on environment and the cardiovascular system. *Circulation*, 63:243A–271A, 1981.

Rom, W. N. (ed), *Environmental and Occupational Medicine*, 2nd edition, Little, Brown and Company, Boston, 1992.

Ross, R., The pathogenesis of atherosclerosis: a perspective for the 1990s. *Nature*, 362:801–809, 1993.

Schoen, F. J. and Gimbrone, M. A., *Cardiovascular Pathology, Clinicopathologic Correlation, and Pathogenetic Mechanisms*, Williams & Wilkins, Baltimore, 1995.

Silver, M. E., *Cardiovascular Pathology*, 2nd edition, Churchill Livingston, New York, 1991.

15

Toxic Responses of the Immune System

Amira F. Gohara

CONTENTS

1-56670-239-9/97/$0.00+$.50
© 1997 by CRC Press, Inc.

INTRODUCTION

The immune system consists of a network of cells and chemical mediators. The main function of this system is to protect the host against infections, tumors, foreign body, etc. The major components of the network are the leukocytes, the immunoglobulins and complement. They all interact with each other to keep the delicate balance of immunity. The functions of the immune system that can be altered as a result of exposures to toxins are phagocytosis, humoral immunity and cellular immunity.

There are four types of toxic reactions of the immune system: immunosuppression, autoimmunity, hypersensitivity or allergy, and neoplasia. This chapter will be divided into three main sections: the normal immune system, lab assessment of the immune system, types of hypersensitivity and effects of toxins on the immune system.

THE NORMAL IMMUNE SYSTEM

To be able to understand the effects of toxins on the immune system, we will briefly review the normal structure and function of the immune system.

Cells of the Immune System

Macrophages

These cells have three main functions:

- Process antigens and present them to various lymphocytes, especially the T cells.
- Secrete chemical mediators that regulate the T and B cell functions.
- Phagocytosis.

Lymphocytes

There are two types of lymphocytes:

- B cells responsible for humoral immunity
- T cells responsible for cellular immunity

Neutrophils

Although these are not an integral part of the immune system, they should be listed here because of their role in phagocytosis through chemotaxis and lysosomal enzymes.

Chemical Mediators of the Immune System

Immunoglobulins

Immunoglobulins are proteins produced by the B cells. These proteins are called antibodies. They are divided into five classes: IgG, IgM, IgA, IgD, IgE. They all have light and heavy chains. They differ from each other in their molecular weight, serum concentration and functions. Only IgG crosses the placenta and only IgG and IgM fixes complement (see below). IgE is associated with allergic reactions. IgA has two components, serum and secretory.

Lymphokines

Substances produced by T cells and subdivided into four groups:

- Mediators interacting with macrophages such as migration inhibition, and migration activation factors.
- Mediators interacting with neutrophils such as chemotaxic factors.
- Mediators interacting with lymphocytes such as transfer factors.
- Others such as osteoclast activating factors.

Complement

The complement system consists of several plasma proteins that are activated via two distinct pathways, "classic" and "alternate". Their activation results in several biological activities ranging from lysis of cells to chemotaxis and mediation of inflammation.

LAB ASSESSMENT OF THE IMMUNE SYSTEM

Phagocytosis

- Cell counts (total WBCs and differential count)
- Chemotaxis test
- Nitroblue tetrazolium test to assess the intracellular oxidative enzymes
- Chemiluminescence to assess the generation of the superoxide radicals
- Bacterial killing amines

Humoral Immunity

- Enumeration of B cells either by immunofluorescence or flow cytometry and monoclonal antibodies to surface markers
- Total protein and protein electrophoresis
- Assessment of immunoglobulin levels
- Immunoelectrophoresis

Cellular Immunity

- Enumeration of T cells either by using rosettes or flow cytometry and monoclonal antibodies to assess not only total T cells, but also T cell subsets.
- T cell function studies (ConA and PHA stimulation tests)
- Skin testing

Assessment of Complement

- Total hemolytic complement (CH_{50})
- Assessment of complement components C_3, C_4, C_2, etc.

The aforementioned lists highlight only the major tests used in the assessment of the immune system.

HYPERSENSITIVITY

The adverse responses of the immune system to environmental toxins can be divided into four different groups. Type I (anaphylactic reactions) mediated by IgE, mast cells and basophils are usually associated with histamine and other vasoactive amines release. Such reactions are hay fever, allergic rhinitis, ataxic dermatitis, etc. Type II (cytolytic reaction) are mediated by complement fixing antibodies and result in cell destruction. Type II reactions are the hemolytic anemias. Type III (immune complex reactions) are circulation antigen antibody complexes deposited in various tissues, fix complement and result in tissue destruction. Such diseases as systemic lupus and other autoimmune diseases are in this group. Type IV are delayed hypersensitivity reactions mediated by T cells.

EFFECTS OF TOXINS ON THE IMMUNE SYSTEM

Since this is an extensive subject, five major types of toxic reactions in the immune system will be discussed here with illustrative examples of the specific effects.

Allergic Reactions Induced by Toxins

Each one of us is exposed daily at work as well as at home to a variety of chemicals capable of inducing an allergic reaction. Such reactions could elicit asthma, rhinitis, dermatitis, gastroenteritis, etc. The major chemicals known to induce an allergic response in sensitive individuals are as follows:

Metals.   Sensitive hosts develop dermatitis and eczema from contact to various types of metals including nickel used in jewelry as well as zippers and chromium used in leather tanning. Medicinal gold salts can also result in a hypersensitivity reaction. Mercury present in films can also cause allergies.

Detergents and formaldehyde.   These chemicals used in textile finishes can also result in a contact type dermatitis.

Resins.   Resins can be hazardous to sensitive individuals when exposure is high. Plastics and resins can cause asthma and chronic cough. Resins are present in glues and various paints among other sources, exposure to resins can cause respiratory problems in sensitive individuals

Cosmetics and enzymes.   Cosmetics and enzymes in detergents sometimes result in allergic dermatitis.

Immunosuppression

Several medicinal as well as drugs of abuse and environmental pollutants have been found to have an immunosuppressive effect. A few examples of immunosuppresive medicinal drugs are as follows:

Steroids.   Lymphopenia with suppression of both T and B lymphocytes as well as neutropenia follows the administration of steroids. Steroids also suppress phagocytosis and prevent the release of liposomal enzymes from the neutrophils.

Cytotoxic drugs.   Cytoxan and azerthiopine suppress both T and B cells and humoral immunity and increase susceptibility to infection. Phagocytosis is also suppressed.

Cyclosporine A.   This drug used to prevent rejection of transplanted organs has several side effects on the immune system. It suppresses T helper cells and increases susceptibility to viral and fungal infections. Cyclosporine also induces malignant lymphomas.

Drugs of abuse.   The relationship between drugs of abuse and immunosuppression is still very controversial and research is still needed. A few examples of drugs of abuse are:

- *Alcohol*. All facets of the immune system seem to be affected by chronic alcohol consumption. Alcoholics have been found to have hypergammaglobulinemia, low T cell counts, abnormal T cell functions as well as depressed chemotaxis. The exact mechanism of action of alcohol on the immune system is not clearly understood.
- *Cocaine*. There is no clear evidence that cocaine has any substantial effect on the immune system.

- *Nitrites*. With the AIDS crisis, nitrites received a great deal of publicity as immunosuppressive agents because they are heavily used by homosexuals. However, several researchers were unable to demonstrate a clear correlation between nitrites and immunosuppression, therefore, further studies are needed in this area.
- *Cannabis*. This has been found to suppress all aspects of the immune system in animal models.
- *Heroin*. This appears to have an effect similar to chronic alcoholism resulting in suppression of T cell counts as well as T cell functions and hypergammaglobulinemia which interestingly enough is very similar to the immunologic changes seen in AIDS patients.

It appears from all the studies to date that the correlation of drugs of abuse and immunosuppression remains an area open for further studies to resolve unanswered issues and questions.

Environmental Pollutants

Pesticides. Pesticides, especially organophosphates, have been found to depress antibody responses in animal models. On the other hand, carbamate insecticides (e.g., serum) have been associated with suppressed antibody response as well as altered phagocytosis.

Urethane. This has been associated with suppression of the natural killer cell activity.

Aromatic hydrocarbons. Used as flame retardants, plasticizers, etc., these are associated with atrophic changes of the lymphoid organs, as well as lower levels of circulating immunoglobulins.

Benzene. Workers exposed to benzene have been found to have decreased serum complement, as well as immunoglobulins.

Metals. Lead decreases both cellular and humoral immunity. Mercury suppresses PHA response.

Ozone. There is an increase of respiratory diseases in individuals exposed to high concentrations of ozone which appears to be caused by decreased resistance to infection as a result of altered phagocytic and bactericidal activity of pulmonary macrophages.

Autoimmunity

A large number of drugs have been associated with a variety of autoimmune phenomena.

The mechanisms by which drugs induce autoimmunity are multiple. Release of segmented autoantigens, cross reactions of chemicals with autoantigens as well as direct effects on T cells are but a few of the mechanisms implicated in toxic autoimmunity.

Antibiotics.   Penicillin can cause an autoimmune hemolytic anemia which disappears upon cessation of therapy. Methicillin produces an autoimmune-induced acute interstitial nephritis with circulating antibodies to tubular epithelium. In the hemolytic anemia mentioned above, penicillin acts as a hapten. Nitrofurantoin produces a lupus-like nephritis.

Methyldopa (Aldomet).   This drug is associated with an autoimmune hemolytic anemia.

Hydralazine.   This is known to cause a systemic lupus-like syndrome associated with arthritis, fever, skin rashes, myalgia and a positive ANA test.

Phenytoin.   This drug can initiate an SLE-like syndrome associated with hemolytic anemia, necrotizing angiitis, skin rash and positive ANA and Coomb's test.

Metals.   Autoimmune diseases are induced by chronic administration of mercury especially a Goodpasture-like syndrome with circulating anti-GBM.

Neoplastic Changes

It is well known environmental factors play a major role in the initiation of human tumors. These factors vary from social habits, like smoking, alcohol abuse, etc. to pollutants present in air and water. This is especially true in various industries. The major ones are:

Asbestos.   Asbestos causes mesothelioma in humans. Some of the occupations associated with asbestos exposure include insulation work, auto brake industry and some textiles.

Aromatic hydrocarbons.   These are associated with sarcoma of the lung.

Benzidine.   This is associated with bladder and liver malignancies. Industries associated with benzidine exposure include dye and plastic workers.

Vinyl chloride.   This is associated with hepatocellular carcinoma exposure in rubber workers.

These are just a few of the industrial chemicals associated with the initiation of neoplastic diseases.

REFERENCES

Dean, J. H., Luster, M. I., Munson, A. E., and Kimber, I., *Immunotoxicology and Immunopharmacology*, 2nd edition, Raven Press, New York, 1994.

Descortes, J., *Immunotoxicology of Drugs and Chemicals*, Elsevier, Amsterdam, 1986.

Sirois, P. and Rola-Pleszcynski, M. (eds), *Immunopharmacology*, Elsevier, Amsterdam, 1982.

Smialowicz, R. and Holsapple, M. P., *Experimental Immunotoxicology*, CRC Press, Boca Raton, FL, 1995.

Part III: The Work Environment

16

Chemical Exposure and Health Risk Assessment

Stephen K. Hall

CONTENTS

INTRODUCTION

Under specific circumstances, health risk can be a probability of injury, disease, or even death. All exposures to chemicals carry some degree of health risk. Many health risks are understood with a relatively high degree of accuracy. For example, exposure of the skin to a corrosive chemical such as sulfuric acid will result in a practical certainty that injury will occur. On the other hand, the health risks associated with many chemical exposures cannot be assessed readily. Assessment of the health risks of chemical exposures that do not cause immediately observable forms of injury or disease is a very involved task.

Health risks may be expressed in quantitative terms, taking values from one to zero. A risk of one is the practical certainty that injury will occur. Zero risk is the practical certainty that injury will not occur. At the present time, science can identify the conditions under which health risks are so low they would generally be considered to be of no practical consequence. In the context of health risk assessment, the term "safe" is not particularly helpful, because it is not possible to identify the conditions under which a given chemical

1-56670-239-9/97/$0.00+$.50
© 1997 by CRC Press, Inc.

exposure is likely to be without risk. Almost all chemicals can be made to produce toxic response under specific conditions of exposure. In this sense, all chemicals are potentially toxic. The important question is not simply that of toxicity, but rather that of risk, i.e., what is the probability that the toxic properties of a chemical will be expressed under actual conditions of exposure? The established "safe" chemical exposure levels are probably risk-free, but science today simply has no tools to prove the existence of a negative, i.e., the absence of risk. In many health risk cases, the risk can only be described qualitatively by the assessor as high, average, low, trivial, or negligible.

HEALTH RISK ASSESSMENT

In recent years, the need to quantify the health risks associated with exposure to chemicals has generated a new interdisciplinary methodology known as health risk assessment. According to the definitions developed by the National Academy of Sciences, health risk assessment is the qualitative or quantitative characterization of the probability of potentially adverse health effects from human exposure to environmental hazards. The outputs of health risk assessment are necessary for informed regulatory decisions regarding worker exposures, plant emissions and effluents, ambient air and water exposures, chemical residues in foods, waste disposal sites, consumer products, and naturally occurring contaminants.

Regulatory actions are based on two distinct processes: health risk assessment and risk management. Risk management is the process of evaluating alternative regulatory options and selecting among them the most appropriate action based upon the results of health risk assessment, available control technology, cost benefit analysis, acceptable risk, acceptable number of cases, policy analysis, and social and political factors. A health risk assessment may be one of the bases of risk management.

The major impetus for conducting health risk assessments comes from federal legislation. Four federal agencies—the Environmental Protection Agency (EPA), Occupational Safety and Health Administration (OSHA), Food and Drug Administration (FDA), and Consumer Product Safety Commission (CPSC)—have been given primary authority to regulate activities and substances that present chronic health risks.

According to the National Academy of Sciences, there are four steps in every health risk assessment: (1) health hazard identification; (2) dose-response assessment; (3) exposure assessment; and (4) risk characterization. Each of these steps is discussed below.

HEALTH HAZARD IDENTIFICATION

The health hazard identification step in health risk assessment involves collecting, organizing, and evaluating toxicity data on the types of health injury or disease that may be produced by the chemicals and on the conditions of exposure under which injury or disease is produced. It may also involve characterization of the behavior of a chemical within the body and the interactions it undergoes with organs, cells, or even parts of organs or cells. Data of the latter types are of value in answering the ultimate question of whether the forms of toxicity known to be produced in experimental animals are also likely to be produced in the human population. In other words, this step is conducted to determine whether or not, and to what extent, it is scientifically correct to infer that toxic effects observed in one setting will occur in another setting.

It is important to note that the amount and type of information on the toxicity of a chemical can be a major source of uncertainty about its potential for toxic effects in humans. If a chemical has been widely used for an extended period of time, toxicologists can be relatively sure in their predictions of the chemical's potential to cause injurious effects. In many cases, however, the data are more limited, and the types of tests that have been performed may not have detected all aspects of the chemical's potential toxicity.

Epidemiology is the study of the distribution and determinants of diseases and injuries in human populations; it is concerned with the extent and types of illnesses and injuries in groups of people and with the factors that influence their distribution. The epidemiologic approach is a valuable methodology for assessing the association of chemical exposure and occurrence of a disease in a human population. It is especially important in the regulatory process because the results are necessary to elucidate the risk of human chemical exposure incurred by human beings without the uncertainty of interspecies extrapolation. Epidemiologic studies may be either descriptive or analytical. Descriptive studies may include surveys and anecdotal reports to suggest hypotheses, while analytical studies utilize retrospective and prospective studies to test the hypothesis.

In reality, most toxicity data comes from animal studies. Accordingly, assumptions must be made about the applicability of toxicity test conditions such as route of entry and dosing, and about the biological similarity of various mammalian species. Unless there are specific data on human toxicity that refute a specific finding of toxicity in animals, or unless there are biological reasons to consider certain types of animal data irrelevant to humans, it is generally assumed that human toxicity can be inferred from observations in animal studies.

Generally, the assumption that all forms of toxicity observed in animal studies will also be observed in human is a prudent one and has been accepted for public health policy. However, most toxicologists agree that animal data alone may not be considered sufficient to establish the cause of identical effects of a specific chemical to expose humans under similar conditions.

For carcinogens, it has been demonstrated that almost all known human carcinogens are carcinogenic in some but not all, laboratory animal species. However, the converse relation is not true. It has not been demonstrated that all animal carcinogens are human carcinogens. In fact, evidence that differences between humans and animal species in susceptibility to carcinogens are likely to stem from finding of qualitative differences in carcinogenic response among different laboratory animal species. Recognizing that the nature of the evidence suggesting carcinogenic potential varies widely among chemicals, the Environmental Protection Agency, and the National Toxicology Program have categorized different carcinogens according to the strength of the underlying evidence.

A mutagenic effect is any alteration in the genetic material of an individual or species, usually one that results in a permanent hereditable change, as a result of exposure to an agent such as a carcinogen or ionizing radiation. A mutagen is a chemical that can induce alterations in the genetic material of either somatic or germinal cells. Mutations carried in germ cells are heritable and may contribute to genetic disease, whereas mutations occurring in somatic cells may be implicated in the etiology of cancer. There are many test systems currently available that can contribute information about the mutagenic potential of a chemical, and there is also considerable experimental evidence that supports the proposition that most chemical carcinogens are mutagens and that many mutagens are carcinogens. As a result, a positive response in a mutagenicity assay is supportive evidence that the agent tested is likely to be carcinogenic. In the absence of a positive animal bioassay, however, such data are not sufficient to support a conclusion that an agent is carcinogenic. Because they are inexpensive and rapid, short-term mutagenicity assay tests are valuable for screening chemicals for potential carcinogenicity and lending additional support to observations from epidemiologic and animal studies.

Comparison of a chemical's molecular structure or physical properties with those of known carcinogens provides some evidence of potential carcinogenicity. Experimental animal studies and short-term mutagenicity tests support such associations for a few structural classes. However, studies of molecular structure and activity in carcinogenesis are best used to identify potential carcinogens for further investigation and may be useful in priority setting for carcinogenicity testing. The types of studies that can have input in health hazard identification are summarized in Figure 16.1.

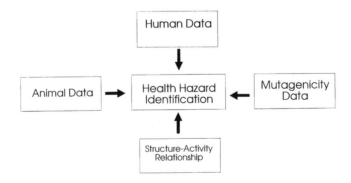

FIGURE 16.1 Health hazard identification.

DOSE-RESPONSE ASSESSMENT

The characteristics of exposure and the spectrum of effects come together in a correlative relationship customarily referred to as the dose-response relationship. This relationship is the most fundamental and pervasive concept in toxicology.

A number of assumptions must be considered before the dose-response relationships can be used (Figure 16.2). The first is that the response is due to the chemical administered. However, to describe the relationship between a toxic chemical and an observed response, one must know with reasonable certainty that the relationship is indeed a causal one. The second assumption is that the response is indeed related to the dose. This sounds simple and obvious. In reality, this assumption is a composite of three concepts: (1) the response is a function of the concentration at the reactive site; (2) the concentration at the reactive site is a function of the dose; and (3) response and dose are causally related. The third assumption is that one has both a quantifiable method of measuring and a precise means of expressing the toxicity. Although many endpoints are available, they are often indirect measures of toxicity. The ideal endpoint is the one closely associated with the molecular events resulting from exposure to the chemical in question.

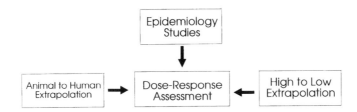

FIGURE 16.2 Dose-response assessment.

In applying toxicity data from animal studies to estimate risks at human exposure conditions, at least two major extrapolations are required: dose extrapolation across species and extrapolation from high to low doses. Interspecies adjustments of dose are necessary to compensate for differences between laboratory animals and humans in size, life span, and basal metabolic rate. Such interspecies adjustments of dose are also called interspecies scaling factors. The most commonly used measures of doses are milligrams of chemical per kilogram of body weight of the animal per day (mg/kg/day), or milligrams of chemical per square meter of body surface area per day (mg/m²/day). Comparison of risk estimates generated from human data and animal data on a number of chemicals indicates that for some, use of body weight provides greater agreement between risk values, while surface area is better for others. There is no general rule as to which scaling factor is more appropriate for a specific chemical.

For any given chemical under specified exposure conditions, the degree of response generally decreases with decreasing dose. In addition, the type of adverse effect may also change with dose. For purposes of health risk assessment, chemicals are usually divided into two categories, carcinogens, and noncarcinogens. Somewhat different methods are used for estimating the risk of exposure to each of these two categories. Noncarcinogenic effects such as skin irritation, organ damage, birth defects and many others are generally thought not to occur until some threshold level of exposure is exceeded. In contrast, many toxicologists consider that carcinogens pose a finite risk at all exposure levels. This "no-threshold" assumption for carcinogens has been adopted by federal agencies as a product practice to protect public health. It should be noted that this assumption may not hold for certain classes of carcinogens because they appear to act through mechanisms that require a threshold dose to be exceeded prior to initiation of the carcinogenic process.

The term "noncarcinogen" is more useful to describe a situation rather than an inherent biological property. It is used to describe either a chemical that has not been shown to be carcinogenic or one that has not been studied for carcinogenic effects. Noncarcinogenic risks are estimated by a somewhat different procedure than carcinogenic risks. Since noncarcinogenic effects are thought to have a threshold dose below which the effect will not occur, the dose-response assessment involves estimation of the threshold dose and determination of the no-observed-effect level (NOEL) from observations in exposed humans or experimental animals. Not all effects of chemical exposure are adverse to health. Thus, the distinction is often made between a NOEL and a no-observed-adverse-effect level (NOAEL). NOELS or NOAELS are divided by safety factors of 10, 100, or 1,000 to obtain an acceptable daily intake (ADI) that is used as a basis for risk characterization.

EXPOSURE ASSESSMENT

Before risk can be characterized from the NOEL for noncarcinogenic chemicals, the magnitude and nature of human exposure and dose must be estimated. Exposure refers to direct contact between the subject and the environmental media which may be air, water, soil or diet, etc., that contain the chemical of interest, or between the subject and the chemical such as an accidental splash onto the skin. A dose results when the chemical of interest is absorbed into the body. The absorbed dose is ordinarily measured as total mass of the chemical absorbed divided by the body weight of the subject and is expressed as milligram of chemical per kilogram of body weight (mg/kg). Sometimes, it is also necessary to specify the time over which a dose is received.

The major routes by which toxic chemicals gain access to the body are through the lungs, skin, gastrointestinal tract, and other parenteral routes. The vehicle, environmental media, and other formulation factors can markedly alter the absorption following exposure. In addition, the route of entry can influence the toxicity of chemicals.

Industrial and environmental exposure to toxic agents most frequently is the result of inhalation and topical exposure. Comparison of the lethal dose of a chemical by different routes of exposure often provides useful information concerning the absorption of the chemical. In instances when the lethal dose after oral or topical administration is similar to the lethal dose for intravenous administration, the assumption is that the toxic chemical is absorbed readily and rapidly. Conversely, in those cases where the lethal dose by the dermal route is higher at several orders of magnitude than the oral lethal dose, it is likely that the skin provides an effective barrier to poisoning by the chemical.

Toxic effects by any route of exposure can also be influenced by the concentration of the chemical in its vehicle, the total volume of the vehicle, and the properties of the vehicle to which the biologic system is exposed, and the rate at which exposure occurs. Studies in which the concentration of the chemical in the blood is determined at various times after exposure are often needed to clarify the role of these and other factors on the toxicity of a chemical.

Toxicologists usually divide the exposure to chemicals into four categories: acute, subacute, subchronic, and chronic. Acute exposure is defined as exposure to a chemical for less than 24 h and to a single administration. Subacute, subchronic, and chronic refer to situations of repeated exposure. Subacute exposure refers to repeated exposure to a chemical for one month or less; subchronic for one to three months, and chronic for more than three months. These three categories of repeated exposure can be by any route, but most often it is by the oral route.

For may chemicals, the toxic effects following a single exposure are quite different from those produced by repeated exposure. For example, the primary acute toxic manifestation of benzene is central nervous system depression, but repeated exposures can result in leukemia. Acute exposure to chemicals that are rapidly absorbed is likely to produce immediate toxic effects, but acute exposure can also produce delayed toxicity that may or may not be similar to the toxic effects of chronic exposure. Conversely, chronic exposure to a toxic chemical may produce some immediate effects after each administration, in addition to the long-term and low-level effects of the chemical. In characterizing the toxicity of a specific chemical, information is needed for the single-dose or acute effects, long-term or chronic effects, as well as exposure effects of intermediate duration.

The other time-related factor that is important in the temporal characterization of exposure is the frequency of exposure. In general, fractionation of the dose reduces the effects. A single dose of a test substance that produces an immediate severe effect might produce less than half the effect when given in two divided doses and no effect when given in ten doses over a long period of time. Such fractionation effects occur when biotransformation or excretion occurs between successive doses or when the injury produced by each exposure is partially or even fully reversed prior to the next exposure. It is obvious that with any type of multiple dose the production of a toxic effect is not only influenced by the frequency of exposure, but may in fact, be totally dependent on frequency rather than duration of exposure.

Chronic toxic effects occur if the chemical accumulates in the biologic system, i.e., absorption exceeds elimination, biotransformation, and/or excretion; if it produces irreversible toxic effects; or if there is insufficient time for the system to recover from the toxic damage within the exposure frequency interval.

Characterizing the exposed population is the final step of exposure assessment. Groups of individuals may be categorized according to the magnitude of the dose they receive. Some may be exposed to several media and by several routes. Others may have more limited exposure. They may also be categorized according to their individual history of past exposure that might have contributed to present injury or disease. Knowledge of past exposure is essential if the toxicity and dose comparisons are to be meaningful. Finally, the number of people exposed is important in some circumstances, e.g., if both individual and population risks are to be taken into account in decision making.

RISK CHARACTERIZATION

Risk characterization is the final step of a health risk assessment (Figure 16.3). The analysis and information of the first three steps are combined to generate statements of risk. For non-carcinogenic or threshold toxicants, a human ADI is estimated by dividing the NOEL by a safety or uncertainty factor, and the human maximum daily dose (MDD) is then compared with the ADI. Different NOELs may be derived for different adverse effects.

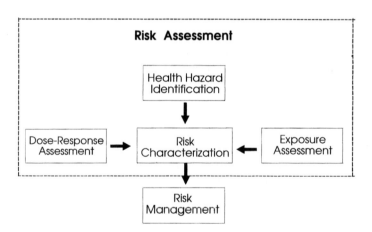

FIGURE 16.3 Risk assessment and risk management.

Two safety factors are commonly used. (1) A 10-fold factor when extrapolating from results obtained in studies involving human exposure: this 10-fold factor is designed to protect the sensitive members of the human population when data have been obtained on "healthy" and "average" individuals. (2) A 100-fold factor when extrapolating from results of long-term animal studies: this represents an additional 10-fold safety factor in extrapolating data from experimental animals to the "average" human. Additional safety factors may be used if data are of questionable quality or if data are from studies of limited duration.

The EPA has recommended uncertainty factors for estimating ADIs for non-carcinogenic chemical pollutants in drinking waters. An uncertainty factor of 10 is used when chronic human data on exposure are available and this is supported by chronic oral toxicity data in animal species. A factor of 100 is used when reliable chronic oral toxicity data are available in some animal species but not in humans. A factor of 1,000 is used when only sub-chronic or limited chronic animal toxicity data are available.

The EPA also recommends an additional uncertainty factor between one and ten in the absence of an available NOEL, when an ADI is estimated from a low-observed-adverse-effect level (LOAEL). This additional factor is incorporated to estimate a NOAEL from the LOAEL. EPA has indicated that in cases where data do not completely fulfill the conditions for a category of uncertainty factors, or appear to be intermediate between two categories, uncertainty factors of intermediate magnitude may be used. In addition, factors smaller than the usual uncertainty factors may be used when there is knowledge of the mechanism of toxicity or of the pharmacokinetics of a chemical in humans or experimental animals.

In characterizing risk for noncarcinogens, the ADI is compared with the MDD that a person might daily sustain for extended period of time. Generally, if the MDD is less than or equal to the ADI, no risk is assumed to exist for almost all members of a population. The MDD may even exceed the ADI without a significant risk arising. As the MDD approaches the experimental NOEL, the risk of toxicity is characterized to be significant. It is important to note that there is no sharp demarcation between "safe" and "unsafe" exposure levels.

Carcinogenic or nonthreshold toxicants, as described earlier, are considered to pose a finite risk at all doses, and the probability of developing cancer increases with increasing dose. The true shape of the dose-response relationship below the experimental dose range cannot be determined from experimental data, since an extremely large number of subjects would be required to detect small responses at very low doses. Three classes of mathematical extrapolation models have been proposed for relating dose and response in the subexperimental dose range: (1) mechanistic models—one-hit, multi-hit, and multi-stage; (2) tolerance distribution models—probit, logit, and Weibull; and (3) time-to-occurrence models—log-normal and Weibull. Other individual models that have been used are: linear, quadratic, and linear-quadratic.

In general, it is impossible to choose among these models based on their fit to experimental dose-response data, since they all usually fit with equal goodness. However, extrapolation below the experimental dose range usually results in divergences of predicted response of several orders of magnitude among the models. The linear and one hit models yield about the same result in the very low dose range and usually are the most conservative in that they yield the highest risk per unit of dose and the lowest acceptable concentration. The linear model consists of linear interpolation between the response observed at the lowest experimental dose and the origin at zero dose and zero response. The overall goal of risk analysis in the subexperimental dose range is not to underestimate human risk. Since the linear model usually yields the highest risk per unit of dose in the very low-dose range, it is unlikely to underestimate risk.

At our present stage of knowledge, the health risks associated with exposure to a chemical cannot be accurately characterized by a single number or even a range of numbers. It is imperative that the quantitative estimates of risk be assessed in conjunction with qualitative factors such as the strength of the evidence that a chemical produces the toxic effects on which the risk was estimated and the uncertainties and assumptions that are inherent in any assessment of health risk. This information is as essential as the quantitative estimation of risk in assessing the true risk and should be considered when actions based on the health risk assessment are considered.

CONCLUSION

In conclusion, health risk assessment is a complex, multifaceted process that is not easily quantified and is currently based on many qualitative decisions. The basic problem is that we can measure the health risks posed by chemicals under specific conditions of exposure but we need knowledge of the risks they may pose under conditions of exposure that fall outside the range of current measurement capabilities.

Quantitative health risk assessment remains the most systematic means currently available to organize, analyze, and present information on environmental chemicals for purposes of deciding to what extent controls on exposure are necessary to protect public health. The most serious potential danger associated with the use of health risk assessment is the failure to recognize its limitations and uncertainties. Risk assessors should be primarily concerned only with scientific issues and leave nonscientific issues such as political and social considerations to risk managers.

REFERENCES

Calabrese, E. J. and Kenyon, E. M., *Air Toxics and Risk Assessment*, Lewis Publishers, Boca Raton, FL, 1991.

Covello, V. T. and Merkhofer, M. W., *Risk Assessment Methods*, Plenum Publishing, New York, 1993.

Fan, A. M. and Chang, L. W., *Toxicology and Risk Assessment*, Marcel Dekker, Inc., New York, 1995.

Hall, S. K., Health risk assessment and chemical exposure, *Pollution Engineering*, 20(12):92–97, 1988.

Hallenbeck, W. H. and Cunningham, K. M., *Quantitative Risk Assessment for Environmental and Occupational Health*, 2nd edition, Lewis Publishers, Boca Raton, FL, 1993.

17

Biotransformation of Occupational Carcinogens

Herman A. J. Schut

CONTENTS

INTRODUCTION

The realization that certain occupations were hazardous to the worker's health by virtue of exposure to chemicals in the workplace had its origins more than 200 years ago. In 1775, Percival Pott, a surgeon at St. Bartholomew's Hospital in London, England, identified the first occupational carcinogen by attributing the high incidence of scrotal cancer in chimney sweepers to their occupational exposure to soot. The development of skin cancer in Northern European workers was similarly found to be correlated with contact of the skin with certain tars and paraffin oils and these findings led in 1907 to the inclusion of such skin cancers in the third schedule of the British Workmen's Compensation Act.

In the latter part of the 19th century it became apparent that there was an association between the development of cancer of the urinary bladder in workers and their prolonged exposure to aromatic amines in an "analine-dye" factory in Frankfurt, West Germany. Subsequent reports from many countries, especially after the first World War showed associations between the occurrence of human bladder cancer and occupations that resulted in exposure to 2-naphthylamine, benzidine, or 4-aminobiphenyl. Since then, numerous similar associations have been noted between the occurrence of human cancer and gross exposure to certain chemicals or chemical mixtures in specific industrial, medical, or societal situations. Chapter 4 discusses the chemicals that are generally recognized as carcinogenic in humans. It should be recognized that the list of chemicals causing cancer in animals is much longer.

Epidemiological studies have shown that, in addition to chemicals, there are many other environmental factors, such as viruses and ultraviolet radiation, that play a role in determining the occurrence of human cancers. It is well known that solar ultraviolet radiation in large doses is responsible for the great majority of skin cancers in light-skinned individuals. Likewise, exposure to ionizing radiation can cause human cancer, but the natural background levels are apparently too low and not variable enough to account for the epidemiological findings. Although there are positive correlations between exposures to certain viruses and certain cancers, current data do not support viral involvement in human cancer as a general situation. As a result, the concept has arisen that certain chemicals in the environment, food, water, air, drugs, and gastrointestinal tract or from social habits, may be important factors in the etiologies of human cancers. Thus, different lifestyles, eating, drinking, and smoking habits, as well as our living, working, and leisure environment may influence exposures to carcinogenic chemicals. It should be noted, however, that environmental factors that modify cancer incidences are not restricted to chemical carcinogens; dietary factors or other agents can modulate the metabolism of chemical carcinogens and/or their reactions with critical targets. They may also influence the selective development and growth of precancerous cells or latent tumors to yield gross tumors. Accordingly, epidemiologic differences between populations in their cancer incidences cannot properly be ascribed to exposures to environmental carcinogens without further detailed studies and understanding of their modulating factors.

INDIRECT-ACTING AND DIRECT-ACTING CARCINOGENS

Much of our knowledge on mechanisms of carcinogenesis has been built up from early studies of only a few organic chemicals. These results have led to the generalization that most carcinogens are really procarcinogens that are metabolized *in vivo* to ultimate carcinogens, frequently via the intermediate formation of proximate carcinogens. The ultimate carcinogens are strongly electrophilic reactants that can combine covalently with cellular molecules which have nucleophilic centers (i.e., negatively charged centers or centers with high electron density such as DNA). Metabolism may also take place to yield nonreactive forms of the procarcinogen (detoxification products) which may be excreted (Figure 17.1). The balance between metabolic activation and inactivation reactions, as well as the inherent carcinogenic activity of the ultimate electrophilic metabolite, are undoubtedly important factors in determining the potency of a chemical carcinogen.

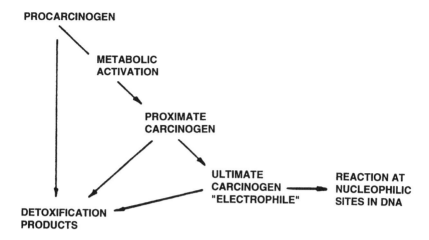

FIGURE 17.1   The metabolic activation and detoxification reactions of chemical carcinogens. (From Tannock, I. F. and Hill, R. P. (eds), *The Basic Science of Oncology,* Pergamon Press, Elmsford, NY, 1987, 94–95. With permission from Pergamon Press.)

It should be noted that there are some organic chemical carcinogens, such as chloroform, thioacetamide, and ethionine that do not fit into this general scheme because they do not appear to bind DNA. Also, the carcinogenic properties of some metals have, for instance, been known for many years, but there is little information on their mechanism of action except for some work on bacterial systems. Even less is known about the action of certain types of asbestos, and most obscure are the mechanisms involved in the induction of cancer by inert mineral oils and plastic films.

The biological properties of electrophilic reactants have been known for many years. Some chemicals of this type, referred to as alkylating agents, have also been used to treat cancer and their mechanism of action has been investigated in detail for the past 30 years. Several carcinogens, as well as most anticancer alkylating agents are direct-acting electrophilic reactants. The chemical structures of some direct-acting carcinogens are shown in Figure 17.2. All behave similarly in that they form electrophilic reactants with generation of the positively charged carbonium ion. This ion can react with numerous biological molecules, such as those containing ionized thiols or acids, or uncharged amino groups. There is some selectivity of action, but in all cases, binding to molecules is widespread and will involve lipids, amino acids, nucleosides, proteins, and nucleic acids. Many binding sites may exist in one molecule. For instance, the nucleophilic sites that may be attacked in DNA are ionized phosphate groups; the N-3, N-7, O^6, C-8 and N^2 positions of guanine; N-1, N-3 and N-7 of adenine; N-3 of cytosine; and O^4 of thymine (Figure 17.3). In proteins, the susceptible regions are the methionyl and cysteinyl groups, the ring nitrogens of histidine, and the C-3 of tyrosine. The main problem in sorting out the mechanism of action of these agents has been in deciding which of these numerous possible covalent reactions contribute to the biological effect. After many years of intensive research, it is now generally accepted that the binding of a carcinogen to DNA is an obligatory prerequisite for the initiation of the carcinogenic process. For several carcinogens, we now know that the persistence of a specific DNA adduct is much better related to a mutagenic cellular event than is the overall extent of covalent binding of the carcinogen. For the most part, the biological consequences of the interaction of chemicals with RNA and proteins are not known, but may contribute to the cytotoxic actions of alkylating agents.

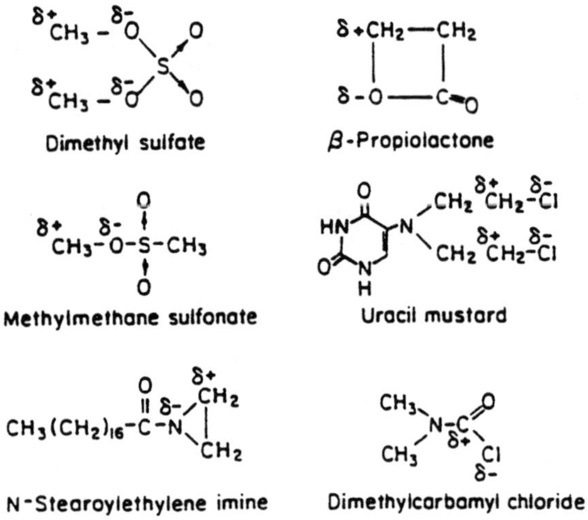

$$\delta^+CH_3 - \delta^-O \underset{\underset{\delta^+CH_3 - \delta^-O}{\overset{S}{\diagdown}}}{\diagup} O$$

Dimethyl sulfate

$$\delta^+CH_2 - CH_2$$
$$\delta^- O - C{\equiv}O$$

β-Propiolactone

$$\delta^+CH_3 - \delta^-O - \overset{O}{\underset{O}{\overset{\parallel}{\underset{\parallel}{S}}}} - CH_3$$

Methylmethane sulfonate

Uracil mustard

$$CH_3(CH_2)_{16} - \overset{O}{\overset{\parallel}{C}} - \delta^- N \diagdown^{\delta^+ CH_2}_{CH_2}$$

N-Stearoylethylene imine

Dimethylcarbamyl chloride

FIGURE 17.2   Chemical structures of direct-acting carcinogens. (From Tannock, I. F. and Hill, R. P. (eds), *The Basic Science of Oncology*, Pergamon Press, Elmsford, NY, 1987, 94–95. With permission from Pergamon Press.)

The remaining discussion in this chapter will pertain to the general mechanism of activation of indirect-acting carcinogens, with some specific examples from major classes of chemical carcinogens.

ENZYMES INVOLVED IN THE METABOLISM OF CARCINOGENS

Enzymes involved in the initial stages of metabolism of chemical carcinogens act to convert foreign, lipophilic compounds into more hydrophilic forms that can be readily excreted. While most metabolites constitute detoxification products, a small proportion are reactive products, a process referred to as enzymatic activation or bioactivation of carcinogens.

The enzyme system that has been implicated in the majority of cases of bioactivation of carcinogens is the microsomal cytochrome P-450 mixed function oxidase (MFO) system. This enzyme system is located predominantly in the endoplasmic reticulum (microsomal fraction) of the cell and, to a lesser

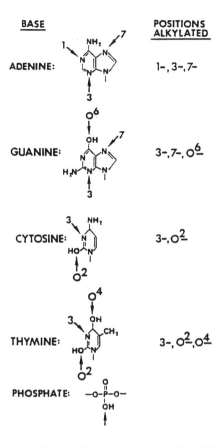

FIGURE 17.3 Sites of alkylation of DNA under physiological conditions. (From Dipple, A. and Bigger, C. A., Mechanisms of action of food-associated polycyclic aromatic hydrocarbon carcinogenesis, *Mutagenesis Research,* 259:263–276, 1991. With permission from Elsevier Science Publishing.)

extent, also in the nucleus. The term P-450 refers to a hemoprotein in the enzyme whose ferrous-carbon monoxide complex exhibits a Soret absorption band near 450 nm and which functions as a MFO or transfering agent. However, P-450 is most often used to describe the enzyme itself. The P-450 enzymes have apparent monomeric molecular weights of 45,000 to 60,000. Hundreds of different isozymic forms occur in each animal species. Each P-450 isozyme exhibits specificity in metabolism of compounds and sterosiomers of a compound can be metabolized at different sites by different P-450s. At least thirteen different P-450s have been purified from rat liver, thirteen from rabbit liver, four from hog liver, and nine from human liver. Many more have been identified by gene cloning but remain unpurified.

The sequence of reactions in the oxidation of carcinogens by cytochrome P-450 is outlined in Figure 17.4. After binding the carcinogen, the enzyme cytochrome P-450 reductase, using NADPH as a donor of reducing equivalents, reduces heme iron from the ferric form to the ferrous form. The heme iron then binds a molecule of oxygen. One atom of oxygen oxidizes the carcinogen and one forms water, concomitantly reoxidizing the iron atom to the ferric state.

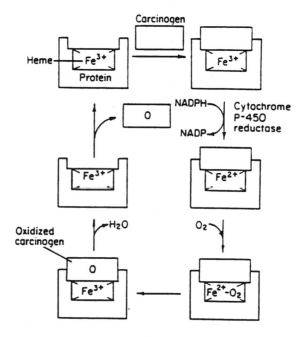

FIGURE 17.4 Sequence of reactions during the oxidation of carcinogens by the cytochrome P-450 system. (From Dipple, A. and Bigger, C. A., Mechanisms of action of food-associated polycyclic aromatic hydrocarbon carcinogenesis, *Mutagenesis Research*, 259:263–276, 1991. With permission from Pergamon Press.)

There are many factors affecting the activity of these oxidative enzymes which in turn can drastically affect the metabolism and potency of a chemical carcinogen. Such factors include the species, strain, sex, age, genetic predisposition, nutritional and hormonal status of the animal, and drugs and other chemicals that act as inducers or inhibitors of these enzymes.

Other enzymes besides the cytochrome P-450s may also activate carcinogens. Prostaglandin synthetase, in the presence of arachidonic acid, has been shown to activate a number of carcinogens by oxidative peroxidation. Furthermore, there are other activating reactions catalyzed by nonoxidative (e.g., reductive and hydrolytic) pathways that include cytosolic enzymes. Conjugation reactions (e.g., addition of glucuronic acid, glutathione, glycine, and sulfate) are particularly important for deactivation of carcinogens, but in some cases (e.g., for aromatic amines and amides) can lead to bioactivation.

While the liver is the principal site of the carcinogen metabolizing enzymes, lower concentrations are also found in almost all tissues with the exception

of skeletal muscle and erythrocytes. Virtually all of the carcinogen-metaboliz-
ing enzymes have natural substrates. The endogenous substrates for the P-
450s, for example, appear to be steroids, fatty acids, fat-soluble vitamins,
prostaglandins, leukotrienes, and thromboxanes. The ability of a target organ
to activate carcinogens appears in a number of cases to be a necessary, though
not sufficient, condition to account for the organ specificity of a carcinogen.

Finally, while most of the metabolic pathways of carcinogen metabolism
have initially been worked out in animals, it is now known that the same
pathways, at least qualitatively, operate in the human.

THE METABOLISM OF
POLYCYCLIC AROMATIC HYDROCARBONS

Polycyclic aromatic hydrocarbons (PAHs) consist of three or more fused
benzene rings in linear, angular, or cluster arrangements and contain only
carbon and hydrogen. These compounds are ubiquitously present in our environment, soil, air, food, water, and working domains. Most PAHs found in the
environment probably arise as a result of pollution via airborne particulate
matter from coal burning for heat and energy production, refuse burning, coke
production, and motor vehicle exhaust; in addition, pollution of the earth's
water by PAHs occurs from spillage from industrial wastes and accidental
contamination during shipment of fuels by sea.

The structures of the most common PAHs are given in Figure 17.5. Of
these, benzo(a)pyrene is the most ubiquitous. One survey has indicated that
as much as 894 tons of benzo(a)pyrene are emitted into the air per year in the
United States. Coal burning may contribute as much as 80% to this total
amount, while forest and agricultural refuse burning contribute only 1%.

PAHs have also been detected in various foodstuffs, both cooked and
uncooked. Also, tobacco smoke contains various PAHs including
benzo(a)pyrene, chrysene, benzo(a)anthracene, and dibenz(a,h)anthracene.
The vast majority of all of the PAHs tested in animal bioassays have proven
to be potent carcinogens.

The metabolism of benzo(a)pyrene and benzo(a)anthracene is shown in
Figures 17.6 and 17.7, respectively. The general pathways of metabolism are
similar for the two compounds. The procarcinogen is first oxidized by cytochrome P-450 to epoxides that are located at various positions in the molecule.
P-450 also catalyzes the formation of various quinones and phenols, not all
of which are shown. Some of the epoxides are substrates for conjugation with
glutathione and these products are excreted as a detoxification process. Several
of the epoxides are converted to dihydrodiols by epoxide hydrases which are
enzymes also present in the microsomal fraction of the cells. Dihydrodiols are
further metabolized by cytochrome P-450, yielding dihydrodiol epoxides
which are the ultimate electrophilic forms of PAHs. For benzo(a)pyrene several
such dihydrodiol epoxides are formed (Figure 17.6) but only one yields a

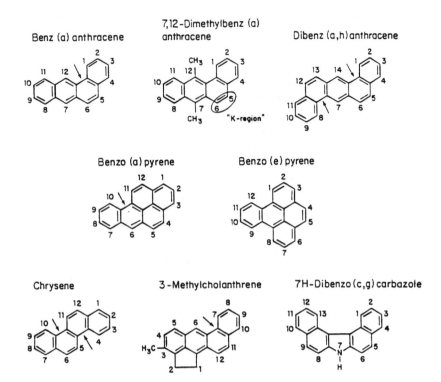

FIGURE 17.5 Structural formulas of polycyclic aromatic hydrocarbons. The arrows indicate the bay regions in the molecule. The "K"-region, shown for example in 7,12-dimethylbenz(a)anthracene, is encircled. (From Zedeck, M. S., Polycyclic aromatic hydrocarbons, a review, *Journal of Environmental Pathology and Toxicology*, 3:537–567, 1980. With permission.)

covalent adduct with DNA, the others being converted to tetrols. All of the reactions shown in Figure 17.6 proceed with a high degree of stereoselectivity.

Initially, much attention was devoted to the so-called K-region epoxides which are formed at a chemically reactive region of high electron density in the molecule (e.g., the 4,5-position of benzo(a)pyrene) (Figure 17.6). Such epoxides can react with nucleophiles and, hence, are candidates for being the actual, ultimate carcinogenic metabolite. K-region epoxides proved to be mutagenic and able to transform cells in culture. Their carcinogenic potency in animals, however, was shown to be less than that of the parent hydrocarbons, raising doubts about their role as carcinogenic forms. When DNA-benzo(a)pyrene adducts were analyzed, it became clear that these were not derived from K-region epoxides. We now know that the activation reactions take place in the "bay region" of the hydrocarbon that is formed by angular rings (Figure 17.5). Such bay-region epoxides have high electrophilic reactivity, are highly mutagenic, are considerably more carcinogenic than the parent hydrocarbon, and account for the major DNA adducts formed *in vivo* from the parent hydrocarbons.

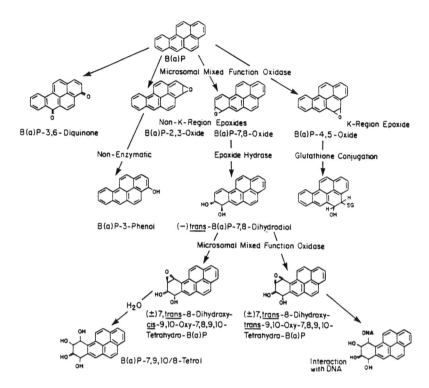

FIGURE 17.6 The metabolism of benzo(a)pyrene [B(a)P]. The metabolic pathways shown are representative of the types of reactions involved in the biotransformation of B(a)P. For clarity, the formation of all the possible epoxide, phenol, quinone, dihydrodiol and tetrol derivatives is not included. (From Zedeck, M. S., Polycyclic aromatic hydrocarbons, a review, *Journal of Environmental Pathology and Toxicology*, 3:537–567, 1980. With permission.)

THE METABOLISM OF AROMATIC AMINES

Aromatic amines and amides comprise a class of compounds widely distributed in industry and research. As pointed out above, several compounds in this class were responsible for the occurrence of bladder cancer in workers in certain industries. The structures of 1-naphthylamine, 2-naphthylamine, benzidine, 4-aminobiphenyl, N,N-dimethyl-4-aminoazobenzene, N-methyl-4-aminoazobenzene, and 2-acetylaminofluorene are shown in Figure 17.8.

The metabolism of 2-acetylaminofluorene (AAF) is outlined in Figure 17.9. AAF can undergo both ring and N-hydroxylation in a number of species, including the human, but the N-hydroxylation step is of unique importance in the activation of AAF to its ultimate carcinogenic form. The ring hydroxylations are considered detoxification reactions. N-hydroxy-AAF is a substrate for conjugation reactions, and its O-glucuronide is quantitatively the most abundant form in most species, with lesser amounts of the O-glucuronide of N-hydroxy-2-aminofluorene being produced. Sulfation of N-hydroxy-AAF has

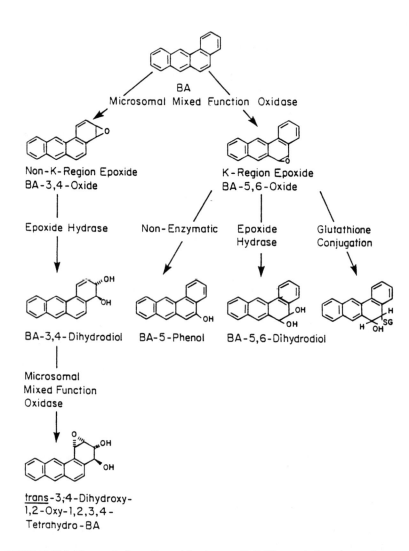

FIGURE 17.7   The metabolism of benzo(a)anthracene (BA). The metabolic pathways shown are
representative of the types of reactions involved in the biotransformation of BA.
For clarity, the formation of all the possible epoxide, phenol, and dihydrodiol
derivatives was not included. (From Zedeck, M. S., Polycyclic aromatic hydrocarbons, a review, *Journal of Environmental Pathology and Toxicology*,
3:537–567, 1980. With permission.)

been related to susceptibility of rats to AAF-induced heptocarcinogenesis. The
sulfate ester of N-hydroxy-AAF is a highly unstable compound which reacts
rapidly with nucleic acids and proteins at neutral pH. Hence, there is only
indirect evidence that this conjugate is a proximate carcinogenic form of AAF.
N-hydroxy-AAF is readily deacetylated by microsomal esterases to yield the

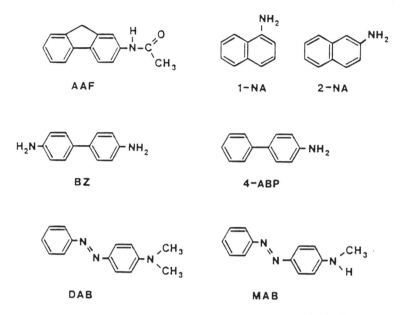

FIGURE 17.8 Structures of carcinogenic aromatic amines. AAF, 2-acetylaminofluorene; 1-NA, 1-naphthylamine; 2-NA, 2-naphthylamine; 4-ABP, 4-aminobiphenyl; BZ, benzidine; DAB, N,N-dimethyl-4-aminoazobenzene; MAB, N-methyl-4-aminoazobenzene. (From Schutt, H. A. J. and Castonguay, A., Metabolism of carcinogenic amino derivatives in various species and DNA alkylation by their metabolites, *Drug Metabolism Review*, 15:753–839, 1984. With permission from Marcel Dekker.)

hydroxylamine, N-hydroxy-2-aminofluorene. The acetyl group of N-hydroxy-AAF can also be transferred to the hydroxyl moiety, a reaction catalyzed by a soluble N-O-acetyltransferase. A variety of studies have shown that deacetylation or N-O-transacetylation are probably more important than sulfation in the activation of N-hydroxy-AAF to its ultimate carcinogenic form. Whether N-hydroxy-AAF interacts with nucleic acid through the formation of an arylnitrenium ion, or through one-electron oxidation yielding nitroxyl free radicals which dismutate to give 2-nitrosofluorene and N-acetoxy-AAF is not clear. The bulk of the evidence favors the nitrenium ion pathway.

The metabolism of 1-naphthylamine, 2-naphthylamine, and 4-aminobiphenyl is similar to that of AAF in that both ring and N-hydroxylations take place (Figure 17.10) and that N-hydroxylation is required for initiation of the carcinogenic process. After the formation of the N-hydroxylamine, this intermediate is conjugated to the N-glucuronide in the liver, enters the circulation and is excreted in the urine. Under mildly acidic conditions, such as those existing in human and dog urine, the glucuronide is hydrolyzed to the free hydroxylamine, which is readily absorbed by cells lining the bladder. In these, the hydroxylamine may be converted to the electrophillic arylnitrenium ion and may bind to cellular macromolecules and induce cancer.

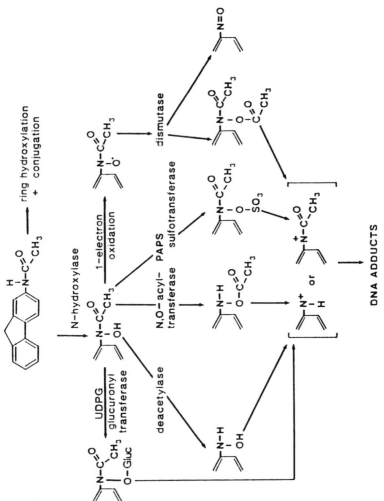

FIGURE 17.9 Metabolic pathways of 2-acetylaminofluorene (AAF). UPDG, uridine diphosphate glucuronic acid; PAPS, 3'-phosphoadenosine-5'-phosphosulfate. (From Schutt, H. A. J. and Castonguay, A., Metabolism of carcinogenic amino derivatives in various species and DNA alkylation by their metabolites, *Drug Metabolism Review*, 15:753–839, 1984. With permission from Marcel Dekker.)

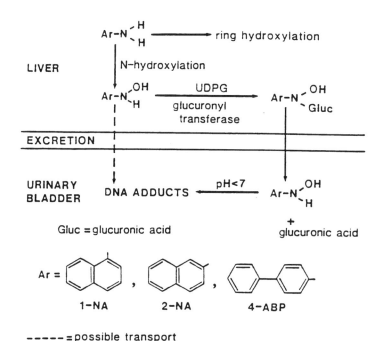

FIGURE 17.10 Metabolic pathways of 1-naphthylamine (1-NA), 2-naphthylamine (2-NA), and 4-aminobiphenyl (4-ABP). UDPG, uridine diphosphate glucuronic acid. (From Schutt, H. A. J. and Castonguay, A., Metabolism of carcinogenic amino derivatives in various species and DNA alkylation by their metabolites, *Drug Metabolism Review*, 15:753–839, 1984. With permission from Marcel Dekker.)

THE METABOLISM OF N-NITROSAMINES

N-Nitrosamines have been shown to be carcinogenic in 39 species of experimental animals. None of the species tested so far has been resistant to their carcinogenic effects. It is therefore unlikely that man would be resistant to the carcinogenic effect of *N*-nitrosamines. Nevertheless, a causal relationship between *N*-nitrosamine exposure and human cancer has yet to be demonstrated. Epidemiological studies have shown that machinists may be exposed to significant amounts of *N*-nitrosoamines present in cutting fluids. Workers in the American rubber industry have a higher incidence of several types of cancer, and relatively high amounts of *N*-nitrosomorpholine have been detected in the air in rubber factories. The etiological role of these *N*-nitrosamines in the increased incidence of cancer in rubber factories is difficult to assess because of the presence of other known human carcinogens.

Studies on the etiology of oral cancer among snuff-dippers could shed some light on the genesis of human cancer initiated by *N*-nitrosamines and might eventually provide direct evidence that *N*-nitrosamines are human carcinogens. Epidemiological studies have linked snuff-dipping to increased risk of oral cancer. *N*-nitrosamines are the only known animal carcinogens which have been detected in significant amounts in snuff tobacco.

The metabolism of dimethylnitrosamine, diethylnitrosamine, and their analogs follows the common pathway illustrated in Figure 17.11. Alpha-Carbon hydroxylation, which is mediated by a cytochrome P-450 MFO, is a crucial step in the bioactivation of *N*-nitrosamines to ultimate carcinogenic forms. The resulting alpha-hydroxy-*N*-nitrosamine (4) has a very short half-life under physiological conditions and rearranges to an aldehyde (5) and primary *N*-nitrosamine (6). The latter can tautomerize to a diazohydroxide (7). In the cytosol, this diazohydroxide is in equilibrium with the very unstable alkyldiazonium ion (8), which can react directly with cellular nucleophiles or lose nitrogen to yield a carbonium ion (9). The latter can, under physiological conditions, alkylate DNA in various positions (Figure 17.3).

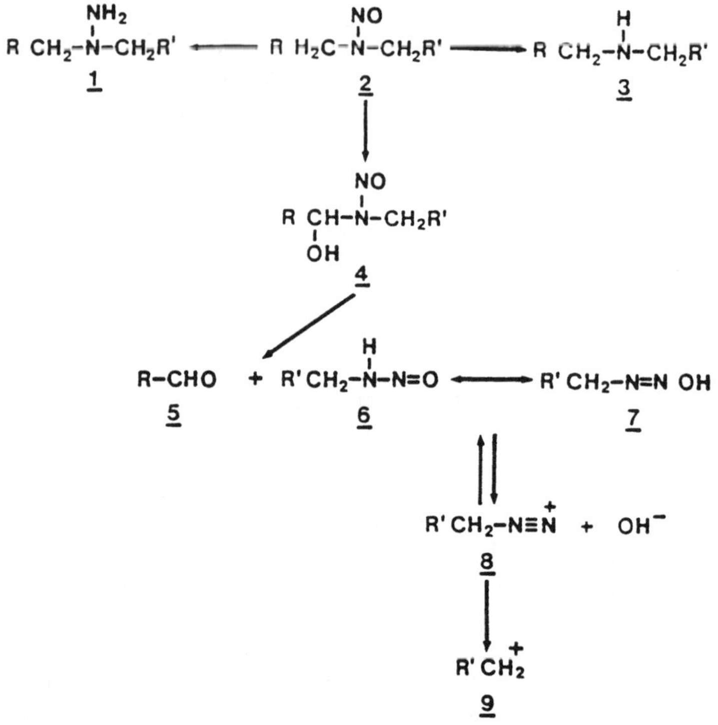

FIGURE 17.11 Metabolic pathways of dimethylnitrosamine (DMN, R = R' = H), diethylnitrosamine (DEN, R = R' = CH_3) and methylethylnitrosamine (R = H,R' = CH_3). (From Schutt, H. A. J. and Castonguay, A., Metabolism of carcinogenic amino derivatives in various species and DNA alkylation by their metabolites, *Drug Metabolism Review*, 15:753–839, 1984. With permission from Marcel Dekker.)

METABOLISM OF OTHER CARCINOGENS

In the above discussion, examples of the metabolism of certain classes of chemical carcinogens or environmental toxins have been presented. Metabolic pathways for many other carcinogens are known but are not included here because of space limitations. For instance, the metabolism of nitroaromatics (including nitrobenzenes, nitrotoluenes, nitrofluorenes, and nitropolycyclic aromatic hydrocarbons), many of which are important intermediates in the production of plastics, rubber chemicals, and dyes, is known. Similarly, the metabolism of important industrial chemicals like vinyl chloride and trichloroethylene has been studied extensively. The metabolism of naturally occurring carcinogens, especially those of plant origin, has received much attention recently because of the human health implications. Such naturally occurring carcinogens include mycotoxins (aflatoxins and sterigmatocystins), pyrrolizidine alkaloids, cycasin, safrole, and tannic acid.

During the last 10 to 15 years a new class of carcinogens, the so-called heterocyclic amines has been discovered. These compounds are formed upon pyrolysis of amino acids or as a result of Maillard (or browning) reactions during the cooking (e.g., barbecuing or panfrying) of meats, poultry, and fish. Because of their obvious potential risk to human health, the metabolism of these compounds and their interaction with DNA is the subject of intensive research in a number of laboratories.

THE ROLE OF SPECIFIC CARCINOGEN-DNA ADDUCTS

While both carcinogen metabolites and specific carcinogen-DNA adducts may be responsible for the toxicity of a chemical, it is generally believed that only one (or sometimes more than one) specific carcinogen-DNA adduct is responsible for initiation of the carcinogenic process. The formation of specific adducts has been shown for some 100 to 200 carcinogens but the structures of many of these have not been convincingly established. Examples of the main binding sites are shown in Table 17.1. For several carcinogens, it has been established that the major adduct is not responsible for the initiation process, but that a quantitatively less important adduct is responsible. In addition, given that adducts are removed ("repaired") at different rates, it has become clear that persistent adducts are much better correlated with mutational events and carcinogenesis than non-persistent adducts. Finally, although a quantitative relationship has been established between the concentrations of several types of DNA adducts and the appearance (usually much later) of tumors, this area needs more study before generalizations can be made.

TABLE 17.1　Main Nucleic Acid or Nucleoside Adducts of Some Carcinogens *In Vivo* or In Neutral Aqueous Solution *In Vitro*

Carcinogen	Adduct Position
Most methylating and ethylating compounds; anti-cancer agents; aflatoxin; sterigmatocystin	7-guanine
1-Naphthylamine	O^6-guanine
Most PAHs; safrole	N^2-guanine
Most aromatic amines; nitropyrenes	8-guanine
Aldehydes	exocyclic NH_2
Hydrazines	pyrimidines
Ethylnitrosourea	phosphotriester

REFERENCES

Dipple, A. and Bigger, C. A., Mechanisms of action of food-associated polycyclic aromatic hydrocarbon carcinogenesis, *Mutagenesis Research*, 259:263–276, 1991.

Hodgson, E., Bend, D. R., and Philpot, R. M. (eds), *Reviews in Biochemical Toxicology*, Volume 5, Elsevier Science, New York, 1983.

Safe, S., Polychlorinated dibenzo-*p*-dioxins and related compounds: sources, environmental distribution, and risk assessment, *Environmental Carcinogenesis and Exotoxicology Review*, C9:261–302, 1991.

Schutt, H. A. J. and Castonguay, A., Metabolism of carcinogenic amino derivatives in various species and DNA alkylation by their metabolites, *Drug Metabolism Review*, 15:753–839, 1984.

Tannock, I. F. and Hill, R. P. (eds), *The Basic Science of Oncology*, Pergamon Press, Elmsford, NY, 1987, 94–95.

Zedeck, M. S., Polycyclic aromatic hydrocarbons, a review, *Journal of Environmental Pathology and Toxicology*, 3:537–567, 1980.

18

Biological Monitoring
of Metal Exposure

Stephen K. Hall

CONTENTS

1-56670-239-9/97/$0.00+$.50
© 1997 by CRC Press, Inc.

INTRODUCTION

Many metals, especially their ions, play a double role in the physiology of organisms: some are indispensable for normal life, while most of them are toxic at elevated concentrations (i.e., they adversely affect the activity and well-being of living organisms). Recent years have brought an increasing concern for the potential toxic effects of metals, metal ions, inorganic compounds, and organometallic compounds.

Toxic metals mainly belong to the trace elements category and many of them are often not coincidental contaminants but may fulfill essential functions. For example, nickel possesses toxic properties of considerable importance in occupational medicine; yet, it has been shown to be an essential element. Aluminum has been recognized recently as being one of the important toxic metals encountered in clinical medicine and has been shown to have some potential biological function. It seems probable that all metals are toxic.

CHRONIC METAL POISONING

Chronic metal poisoning can occur at home, in the work place, or anywhere in the environment. Chronic lead poisoning at home has been a major public health problem for many years. Lead-based paints represent an important source of excessive lead intake in children. In the work place, lead smelting and refining are particularly hazardous with respect to exposure to lead. Mercury exposure is an important problem in the chloralkaline industry. The presence of high concentrations of arsenic in artesian well water is the cause of chronic environmental arsenic poisoning in southern Taiwan.

ACUTE METAL POISONING

Acute metal poisoning is encountered less commonly than chronic metal poisoning. The exposure is usually accidental and can also occur at home, work place, or environment. Accidental ingestion of arsenic trioxide which is used as an herbicide, or thallium compounds which are used as rat poison, have been reported as the cause of acute metal poisoning. Arsenic has been the most popular homicidal agent. This is undoubtedly due to past publicity in fiction literature. More serious and tragic cases have occurred from accidental severe poisoning with methylmercury which was used as a fungicide in the treatment of seed grain. In some Third World countries, the treated seed grain was used directly in the preparation of bread.

ABSORPTION AND DISTRIBUTION

Toxic metals, especially their ions, once absorbed into the body, will bind tightly to structural or cellular components, and this tendency complicates both the clinical picture and therapeutic approach to metal toxicity. Although metals that are absorbed into the body will nearly always appear in the bloodstream, they are usually rapidly distributed to target organs or tissues where they bind to functional groups such as sulfhydryl, amino, phosphate, carboxylate, hydroxy, and other moieties on biomolecules. The most stable of these complexes involve metal binding to sulfur or nitrogen with oxygen binding to a much lesser extent.

There are three important aspects of metal distribution in the body. First, unlike organic molecules, metals are often not susceptible to metabolic detoxification mechanisms. Second, the relative ease with which metals, their ions, as well as inorganic compounds can circulate among body tissues and fluids, allowing to settle among the tissues to which they are most tightly bound. Third, the propensity of metals to partition quickly out of the blood into tissues make it difficult in the evaluation of body burden.

Blood and urine levels of metal usually reflect recent exposure and correlate best with acute effects. An exception is beryllium, the concentration of which in blood and urine has been only qualitatively related to exposure. Beryllium in urine may be detected for years after removal from exposure. Another exception is urine cadmium where increased metal in urine reflects renal damage related to accumulation of cadmium in the kidney. Speciation of toxic metals in urine may provide diagnostic insights. For example, cadmium metallothionein in urine may be of greater toxicologic significance than cadmium chloride.

It has been known for many years that hair contains metals. This may be considered an excretory mechanism for certain metals. Since metals appear to have no functional role in the hair protein, their content in hair seems to vary with exposure. Hair, however, is not suitable as a dynamic system for evaluating immediate past exposure to metals since the individual hair shafts do not have access throughout their length to any fluid transport system. Since hair grows at the rate of about 1 cm in 30 days, clippings would be expected to reflect a historical exposure at best.

BIOLOGICAL MONITORING PROFILES

The various aspects of metal toxicology are not clearly understood for all metals. As a matter of fact, the body of knowledge concerning the individual metals is extremely spotty. A fairly complete picture emerges only for a few metals, namely, lead, mercury, and cadmium. These metals have been studied most intensively for one reason or another. They serve as models. The following are biological monitoring profiles for metals of industrial significance.

Aluminum

Aluminum does not represent an important health hazard in industry. In a healthy person, aluminum appears to be poorly absorbed by the oral route and probably also through the respiratory tract. While the threat of aluminum exposure to healthy individuals has not been established conclusively, toxicity due to aluminum exposure in exposed individuals with impaired renal function has been widely documented. Observations made in these individuals suggest that the relationship between the concentration of aluminum in blood and urine has not been characterized.

Antimony

Antimony and its compounds are not essential to humans. Environment and nutrition influence antimony concentration in human tissues, causing large differences among individuals. Great variations of antimony content were found in blood, urine, and different tissues.

Arsenic

In industry, workers are usually exposed to inorganic arsenic compounds. Whatever the arsenic compound to which the workers are exposed, the absorbed arsenic is rapidly eliminated in the urine either in the unchanged form or in the methylated forms of monomethylarsonic acid and cacodylic acid (dimethylarsonic acid). The biological monitoring of workers should be carried out by measuring the total amount of arsenic present in urine collected at the end of the shift or at the beginning of the next shift. As some marine organisms may contain very high concentrations of organoarsenicals of negligible toxicity that are also rapidly excreted in urine, workers should be instructed to refrain from eating seafood for 2 or 3 days before urine collection. As in urine, the arsenic concentration in blood reflects mainly recent exposure. Arsenic in hair is an indicator of the amount of inorganic arsenic absorbed during the growth period of the hair.

Beryllium

While the major toxicological effects of beryllium are on the lungs, after absorption, beryllium does not localize in the lungs but it is widely distributed in the body. Beryllium can be determined in blood and urine but at the present time these analyses can only be used as qualitative tests for confirmation of exposure to the metal. It is not known to what extent the concentrations of beryllium in the blood and urine may be influenced by recent exposure and by the amount already stored in the body. Urinary excretion of beryllium may be detected for years after removal from exposure. Blood investigations of workers exposed to beryllium indicate that the beryllium ion is bound to a protein to form a beryllium antigen which induces the cell-mediated beryllium hypersensitivity and can be determined *in vitro* by an increase of lymphoblast transformation and also an inhibition of macrophage migration.

Cadmium

Cadmium is a cumulative toxin and mostly bound to red blood cells in blood. Absorbed cadmium accumulates mainly in the kidney and in the liver, with about half of the body burden found in these two organs. In all tissues, cadmium is bound mainly to metallothionein, a protein of high sulfhydryl content with a capacity for binding cadmium, copper, and zinc. The body burden of smokers is about twice that of non-smokers. Cadmium level in blood appears to reflect mainly recent exposure levels. When the total amount of cadmium absorbed has not yet saturated all the available cadmium binding sites, the cadmium concentration in urine reflects mainly the cadmium level in the body. When integrated exposure has been so high as to cause a saturation of binding sites, cadmium level in urine may be related partly to the body burden and partly to recent exposure. In urine, cadmium is bound mainly to metallothionein. Metallothionein analysis presents an advantage over cadmium metal analysis in that it is not subject to external contamination. A radioimmunoassay has been developed for the determination of metallothionein in urine. The determination of cadmium in hair has been proposed to evaluate past exposure.

Chromium

Chromium absorption depends on the oxidation state of the metal. In the trivalent state, chromium is very poorly absorbed. The toxicity of chromium is attributed to the hexavalent state which is also responsible for the carcinogenicity of the metal. Currently, too few data exists to allow evaluation of blood chromium measurements. Except for chromates, all chemical forms of chromium are rapidly cleared from the blood. The major excretory route for absorbed chromium is in the urine. Chromium concentration in urine can be used as an index of recent exposure to hexavalent soluble chromium compounds.

Cobalt

Cobalt is an essential metal present in vitamin B_{12}. Generally cobalt does not seem to accumulate in a specific target organ. The concentrations of cobalt in blood are highly variable. The high variability may be due to sample contamination. Absorbed cobalt is largely excreted in the urine. Urinary concentrations of cobalt in 24-h urine collection appear to correlate well with occupational exposure.

Copper

Copper is an essential trace element for all vertebrates. Blood concentrations of copper vary widely from city to city. Abnormally high values have been observed from people living in copper smelting vicinities. A sex difference has been observed in copper concentration in the body. Males typically have higher blood concentrations than females but females have higher day-to-day variation. The major excretion route of copper is the bile. Urinary copper is reabsorbed.

Iron

Iron is found in virtually every food. The rate of absorption of iron is inversely related to the state of the body's iron stores and it is a complicated process. Absorbed iron is transported by transferrin, a globulin which is normally one-third saturated. There is diurnal variation in serum iron as well as day-to-day variation. Excessive absorbed iron is excreted in the urine.

Lead

Lead is a cumulative toxin that is absorbed by the lungs and the gastrointestinal tract. In blood, lead is mainly bound to red blood cells. In a steady state situation, lead in blood is considered to be the best indicator of the concentration of lead in soft tissue and hence of recent exposure. However, it does not necessarily correlate with the total body burden. Lead in urine reflects the amount of lead recently absorbed. In the same individual, lead in urine fluctuates with time more than lead in blood. The biological tests for lead exposure are mainly based on the interference of lead with several stages of the heme synthesis pathway. Because of the inhibition of δ-aminolevulinic dehydrase by lead, δ-aminolevulinic acid accumulates in tissue and is excreted in urine. The inhibition of ferro-chelatase in young red blood cells leads to an accumulation of protoporphyrin which exists as zinc protoporphyrin, Other enzymatic methods are also available but not commonly used. Determination of lead in hair is another method that could be used for evaluation of exposure to lead.

Manganese

Manganese is absorbed mainly through the lungs. Gastrointestinal absorption appears to be negligible. In blood, manganese is present mainly in red blood cells. Independent studies have failed to find any significant correlation between manganese exposure and the blood manganese level in workers. Concentrations in red blood cells have been reported to be increased in persons with rheumatoid arthritis. Excretion occurs mainly through the bile. Very small amounts are eliminated in the urine and hair.

Mercury

Mercury absorption depends on the form of the metal. Elemental mercury is well absorbed by inhalation. Inorganic mercury compounds are well absorbed after ingestion. Organic mercury compounds are well absorbed by all routes. Once absorbed, mercury is bound to the sulfhydryl groups of proteins. In blood, inorganic mercury is equally distributed between plasma and red blood cells. Inorganic mercury is excreted mainly in the urine and feces. If exposure to mercury vapor has lasted for at least one year, there is a correlation between the concentrations of mercury in blood or urine and the intensity of recent exposure as well as the occurrence of clinical signs of intoxication. Biological monitoring of organic mercury compounds depends on the organic moiety. Short-chain alkyl compounds, mainly methylmercury, are found mainly in red blood cells and excreted through feces. In this case urinary determination has no practical value. Aryl and alkoxyalkyl mercury compounds liberate inorganic mercury *in vivo* and the concentration of mercury in blood as well as urine is indicative of the exposure intensity.

Nickel

Nickel is not a cumulative toxin. It is absorbed mainly through the respiratory tract but also possibly through the gastrointestinal tract and the skin. Once absorbed, it is excreted rapidly. Excretion occurs predominantly in the urine. Several studies have demonstrated that the concentrations of nickel in plasma and urine are indicators of recent exposure to soluble nickel compounds. While a single blood sample concentration of nickel correlates quite well with air sample concentration, urinary concentration of nickel would only correlate well if a 24-h urine sample is collected.

Selenium

Selenium compounds are well absorbed orally and seem to be well absorbed through the lungs. Under conditions of exposure to high concentrations, a volatile metabolite, dimethylselenide, may be eliminated through the lungs. Blood and urine concentrations of selenium reflect mainly recent exposure. It is important to note that concentration of selenium in blood and urine may vary considerably depending on the dietary intake.

Thallium

Thallium is easily absorbed by any route of exposure. Blood is not a reliable indicator of thallium exposure. Only a small fraction of the body burden of thallium in poisoned subjects is in the plasma. Excretion is predominantly in the urine. Determination of thallium in urine is probably a more reliable indicator of exposure than its determination in blood.

Tin

Absorption of tin depends on the form. Ingested inorganic tin is very poorly absorbed, on the order of about 1%. Organotin compounds are better absorbed from the gastrointestinal tract. Short-chain alkyltin compounds may be absorbed through the skin. Whereas metallic tin and inorganic tin compounds are relatively nontoxic, a number of organotin compounds are highly poisonous. Most absorbed tin is found in the red blood cells. However, large variations from day to day and week to week have been observed; such variations have been attributed to dietary intake of tin. Whereas inorganic tin is excreted in the urine, organotin is excreted in the bile.

Vanadium

Vanadium is absorbed mainly by the pulmonary route. The oral absorption rate appears to be negligible. In blood, absorbed vanadium is bound to plasma transferrin. Currently, the correlation between concentrations of vanadium in blood and exposure level has not been discovered. Absorbed vanadium is rapidly excreted in the urine. The determination of vanadium in urine has been proposed for evaluating recent exposure to the metal.

Zinc

Zinc is an essential metal and available in food. Absorbed zinc is present in the red blood cells, white blood cells, plasma, and serum. Although zinc values in blood serum are generally reliable, serum zinc is lower in women taking oral contraceptives, in pregnant women, and in persons undergoing certain stresses such as infections. Excretion of zinc is largely in the feces. Part of the zinc excreted is reabsorbed.

BIOLOGICAL EXPOSURE INDICES

To date, very few biological tolerance values have been established for metals. The American Conference of Governmental Industrial Hygienists annually publishes *Threshold Limit Values and Biological Exposure Indices*. It should be emphasized that the biological exposure indices are valid only in accordance with their corresponding definitions. These values are not applicable to the general population and apply only to workers.

REFERENCES

Baselt, R. C., *Biological Monitoring Methods for Industrial Chemicals*, Biomedical Publications, Davis, CA, 1980.

Carson, B. L., Ellis III, H. V., and McCann, J. L., *Toxicology and Biological Monitoring of Metals in Humans*, Lewis Publishers, Inc., Chelsea, MI, 1986.

Hall, S. K., Health surveillance of hazardous materials workers, *Pollution Engineering*, 24(9):58–62, 1992.

Hall, S. K., Biological monitoring of metal exposure, *Pollution Engineering*, 21(1):128–131, 1989.

Sheldon, L. et al., *Biological Monitoring Techniques for Human Exposure to Industrial Chemicals*, Noyes Publications, Park Ridge, NJ, 1986.

19

Biological Monitoring of Solvent Exposure

Stephen K. Hall

CONTENTS

INTRODUCTION

Solvent molecules can gain entry into a worker's body by inhalation, skin absorption, or ingestion. They may also gain entry by more than one route concurrently. Traditionally, inhalation has been the major exposure route of concern for the worker. Consequently, industrial hygienists have concentrated on keeping atmospheric levels of chemicals below effect levels known as threshold limit values. These are issued annually by the American Conference of Governmental Industrial Hygienists (ACGIH). Threshold limit values refer to airborne concentrations of substances and represent conditions under which it is believed that nearly all workers may be repeatedly exposed day after day without adverse effect.

Examination of the chemical substances listed by the ACGIH shows that a number of the listed substances also has the "skin" notation, which refers to

1-56670-239-9/97/$0.00+$.50
© 1997 by CRC Press, Inc.

the potential contribution to the overall exposure by the cutaneous route including mucus membranes and eye, either by airborne, or more particularly, by direct contact with the substance. Vehicles can also alter skin absorption. While little quantitative data are available describing absorption of vapors and gases through skin, the consensus of industrial hygiene toxicologists is that substances having a "skin" notation may present an additional problem, particularly if a significant area of the body is exposed to the substance for a long period of time. Hence, measurements of atmospheric concentrations of chemical substances with the "skin" notation will not give an accurate indication of the quantity of chemical that might reach the target organ.

Even if a chemical substance does not have a "skin" notation and even if there exists a relationship between the airborne concentration and the worker by a specific route, one still cannot expect that the determination of the airborne concentration may allow an estimate of the total amount of chemical substance absorbed by the exposed worker. There are differences in worker's personal hygiene habits as well as differences in individual absorption rate of the substance through any port of entry.

BIOTRANSFORMATION OF FOREIGN COMPOUNDS

Most foreign compounds entering the body are subject to biotransformation, or commonly known as metabolic transformation. It is important to note that while biotransformation of foreign compounds often involves detoxification, the same metabolic pathway may actually lead to an increase in the toxicity of some compounds.

One important aspect of biotransformation of foreign compounds is increased excretability, and thus a decrease in toxicity. A major route of excretion is the renal system. The kidney is built in such a way that it can handle electrolytes better than nonelectrolytes. Thus, the higher the degree of ionization at body pH, the more readily the chemicals are excreted by the kidney. Ionization in turn depends on the degree of polarizability of the chemical. Polarizability of a chemical is related to the geometry and molecular size. Asymmetric and large molecular substances are more polarizable than symmetric and small ones. The general mechanism of detoxification is summarized in Figure 19.1.

Biotransformation occurs mainly in the liver by enzymatic reactions although other tissues such as the kidneys or lungs may be involved. These enzymatic reactions occurs in two phases: Phase I reactions involve oxidation, reduction, and hydrolysis; and Phase II reactions involve conjugation or synthetic reaction. The primary function of Phase I reactions is to add or alter the functional groups on the parent molecule. These functional groups then permit the Phase I product to undergo a Phase II reaction. In Phase II, the Phase I product is covalently bonded to an endogenous molecule, producing a conjugate, which results in an increase in hydrophilicity of the foreign compound and hence its excretability. For example, toluene is oxidized in Phase I to form

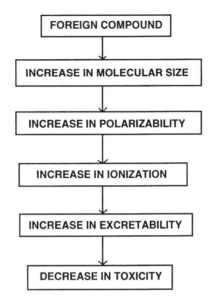

FIGURE 19.1 General mechanism of detoxification.

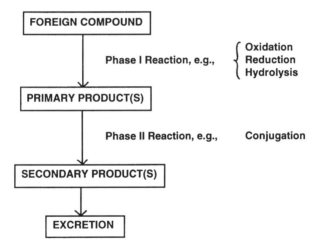

FIGURE 19.2 Metabolic reactions.

benzoic acid which conjugates with glycine in Phase II and is excreted in the urine as hippuric acid. The relationship between Phase I and Phase II reactions is summarized in Figure 19.2.

Oxidation Reactions

Oxidation reactions are essential to animal life. In fact, the enzymatic make-up of living organisms is geared toward oxidizing foodstuff. The lower primary aliphatic alcohols are closely related to carbohydrate and are metabolized to carbon dioxide and water. The higher primary alcohols and glycols are oxidized progressively through the corresponding aldehydes and acids while secondary alcohols are oxidized to ketones. In biological monitoring of solvent exposure, it is necessary for the occupational health professional to be well aware of the metabolic pathway(s) of the foreign chemical in question and in what biological specimen the index chemical can be determined. Biological monitoring of exposure to aliphatic alcohols and glycols is summarized in Table 19.1, aldehydes in Table 19.2, ketones in Table 19.3, and ethers in Table 19.4.

TABLE 19.1   Biological Monitoring of Alcohol and Glycol Exposure

Chemical	Exposure Index	Biologic Specimen
Allyl alcohol	Ally alcohol	Breath
	Acrolein	Blood
Chloroethanol	Chloroacetic acid	Urine
Diethylene glycol	Oxalic acid	Urine
Ethyl alcohol	Ethyl alcohol	Breath
Ethylene glycol	Ethylene glycol	Blood
	Oxalic acid	Urine
Isopropyl alcohol	Acetone	Blood
	Acetone	Urine
Methyl alcohol	Methyl alcohol	Breath
	Methyl alcohol	Urine
	Formic acid	Blood
	Formic acid	Urine

TABLE 19.2   Biological Monitoring of Aldehyde Exposure

Chemical	Exposure Index	Biologic Specimen
Acetaldehyde	Acetaldehyde	Blood
Chloradehyde	Trichloroethanol	Urine
Formaldehyde	Formaldehyde	Blood
	Formic acid	Urine
Furfural	Furoic acid	Urine

While the oxidation of oxygen-containing hydrocarbons such as alcohols and aldehydes is quite straight forward, the oxidation of unsubstituted aliphatic and alicyclic hydrocarbons occurs on a secondary carbon atom next to the primary carbon atom (Table 19.5).

Unsubstituted aromatic hydrocarbons are oxidized to a corresponding aromatic alcohol which is then conjugated with an endogenous chemical and excreted (Table 19.6).

Halogenated hydrocarbons may be excreted unchanged, remain in the living organism unaltered, or oxidized to an alcohol and the corresponding carboxylic acid (Table 19.7).

TABLE 19.3 Biological Monitoring of Ketone Exposure

Chemical	Exposure Index	Biologic Specimen
Acetone	Acetone	Breath
	Acetone	Blood
	Acetone	Urine
	Formic acid	Urine
Methyl ethyl ketone	Methyl ethyl ketone	Breath
	Methyl ethyl ketone	Blood
	Methyl ethyl ketone	Urine
Methyl isobutyl ketone	4-Methyl-2-pentanone	Urine
	4-Hydroxy-4-methyl-2-pentanone	Urine
Methyl n-butyl ketone	2,5-Hexanedione	Urine

TABLE 19.4 Biological Monitoring of Ether Exposure

Chemical	Exposure Index	Biologic Specimen
Diethyl ether	Diethyl ether	Breath
	Diethyl ether	Blood
Dioxane	β-Hydroxyethoxy acetic acid	Urine
Ethylene glycol monoethyl ether	Ethoxyacetic acid	Urine
Ethylene glycol monomethyl ether	Methoxyacetic acid	Urine

TABLE 19.5 Biological Monitoring of Aliphatic and Alicyclic Hydrocarbon Exposure

Chemical	Exposure Index	Biologic Specimen
Cyclohexane	Cyclohexane	Breath
	Cyclohexanol	Blood
	Cyclohexanol	Urine
n-Heptane	*n*-Heptane	Breath
	n-Heptane	Blood
	2-Heptanol	Urine
	2,6-Heptanedione	Urine
n-Hexane	*n*-Hexane	Breath
	n-Hexane	Blood
	2-Hexanol	Urine
	2,6-Heptanedione	Urine
2-Methylpentane	2-Methylpentane	Breath
	2-Methyl-2-pentanol	Urine
3-Methylpentane	3-Methylpentane	Breath
	3-Methyl-2-pentanol	Urine

While aliphatic amines are oxidatively deaminated into the corresponding aldehydes by loss of ammonia, oxidation of aromatic amines may occur either on the amino group or on a secondary carbon atom in the ring (Table 19.8). Aliphatic and aromatic nitro compounds are not further oxidized but many of

TABLE 19.6   Biological Monitoring of Aromatic Hydrocarbon Exposure

Chemical	Exposure Index	Biologic Specimen
Benzene	Benzene	Breath
	Benzene	Blood
	Phenol	Urine
Biphenyl	4-Hydroxybiphenyl	Urine
Ethylbenzene	Ethylbenzene	Blood
	Ethylphenol	Urine
Isopropylbenzene	Dimethylphenylcarbinol	Urine
Naphthalene	α- and β-Naphthol	Urine
Styrene	Styrene	Breath
	Styrene	Blood
	Mandelic acid	Urine
	Phenylglyoxylic acid	Urine
Toluene	Toluene	Breath
	Toluene	Blood
	Hippuric acid	Urine
	o-Cresol	Urine
Xylene	Xylene	Breath
	Xylene	Blood
	Methylhippuric acid	Urine

them produce a physiological effect know as methemoglobinemia which affects oxygen transport in the blood.

Reduction Reactions

Reduction reactions in the metabolism of organic compounds are much less common than oxidation reactions. This is because they go counter to the general trend of biochemical reactions in living tissue. Yet it must be realized that all enzymatic reactions in a living organism are fundamentally reversible. A well-known example of a reduction reaction is the biotransformation of chloral hydrate, the oldest hypnotic, into trichloroethanol. In the case of aromatic polynitro compounds, typically only one of the nitro groups becomes reduced.

Hydrolysis Reaction

Certain organic compounds require cleavage or degradation before they can be further metabolized. The most common reaction of this type is the hydrolysis of an ester to an alcohol and an acid. For example, aspirin is hydrolyzed *in vivo* to acetic acid and salicylic acid, the latter containing an alcohol functional group. Amides, hydrazides, and nitriles may also be hydrolyzed.

Conjugation Reactions

Conjugation reactions appear to have developed chiefly for the metabolism of foreign compounds. This is the reaction that is most successful in terms of

TABLE 19.7 Biological Monitoring of Halogenated Hydrocarbon Exposure

Chemical	Exposure Index	Biologic Specimen
Carbon tetrachloride	Carbon tetrachloride	Breath
	Carbon tetrachloride	Blood
Chloroform	Chloroform	Breath
	Chloroform	Blood
p-Dichlorobenzene	2,5-Dichlorophenol	Urine
Hexachlorobenzene	Hexachlorobenzene	Blood
	(Cutaneous Porphyria)	
Methyl chloroform	Methyl chloroform	Breath
	Methyl chloroform	Blood
	Trichloroethanol	Urine
	Trichloroacetic acid	Urine
Perchloroethylene	Perchloroethylene	Breath
	Perchloroethylene	Blood
Polychlorinated biphenyl	Polychlorinated biphenyl	Blood
1,1,2,2-Tetrachloroethane	Dichloroacetic acid	Urine
	Glycolic acid	Urine
	Trichloroethanol	Urine
	Trichloroacetic acid	Urine
Trichloroethylene	Trichloroethylene	Breath
	Trichloroethylene	Blood
	Trichloroethylene	Urine
	Trichloroacetic acid	Urine
1,1,2-Trichloroethane	1,1,2-Trichloroethane	Breath
	1,1,2-Trichloroethane	Blood
	Chloroacetic acid	Urine
Vinyl chloride	Thioglycolic acid	Urine

detoxification. The typical endogenous compounds involved in conjugation are amino acids (such as glycine) and their derivatives. A few simpler endogenous chemical species (such as acetate and sulfate) are also occasionally involved in conjugation. The conjugation of glycine with benzoic acid to form hippuric acid has been discussed.

BIOLOGICAL MONITORING MEASUREMENTS

Biological monitoring of exposure to industrial chemicals provides occupational health personnel with a tool for assessing a worker's overall exposure to an index chemical, i.e., the chemical of analytical interest through measurement of the appropriate determinant(s) in biological specimens collected from the worker at the specified time. In short, this means evaluation of the internal exposure of the worker to a chemical substance, i.e., the internal dose, by a biological method. The determinant can be the chemical itself or its metabolite(s) or a characteristic biochemical change induced by the chemical. Depending on the chemical and the analyzed biological parameter, the term

TABLE 19.8   Biological Monitoring of Exposure to Nitrogen-Containing Organic Compounds

Chemical	Exposure Index	Biologic Specimen
Acetonitrile	Cyanide	Blood
	Thiocyanate	Urine
Acrylonitrile	Cyanide	Blood
	Thiocyanate	Urine
Aniline	Methemoglobin	Blood
	p-Nitrophenol	Urine
Benzidine	Benzidine	Urine
Dimethyl formamide	Dimethyl formamide	Blood
	N-Methylformamide	Urine
Dinitrobenzene	Methemoglobin	Blood
	Nitroaniline	Blood
Ethylene glycol dinitrate	Ethylene glycol dinitrate	Urine
Nitrobenzene	Methemoglobin	Blood
	p-Nitrophenol	Urine
Nitroglycerine	Nitroglycerine	Blood
Trinitrotoluene	Aminodinitrotoluene	Blood

internal exposure or internal dose may cover different concepts. It may mean the amount of the chemical recently absorbed or the amount of the active species bound to the sites of action.

Biological monitoring takes into consideration absorption by routes other than the lungs. Many industrial chemicals enter the body of a worker by multiple routes. The greatest advantage is that the biological parameter of exposure is more directly related to the adverse health effects than atmospheric measurement. Therefore, it offers a better assessment of the health risk to the worker than ambient air monitoring.

Biological monitoring measurements can be made in exhaled air, urine, blood, or other biological specimens such as hair, milk, saliva, tears, nail, perspiration, adipose tissue, and others. Based on the determinant, the specimen chosen, and the time of sampling, the measurements can indicate the intensity of a recent exposure, an average daily exposure, a chronic cumulative exposure, or even a past exposure.

Exhaled Air Analysis

It has been known for decades that many industrial organic chemicals that are inhaled or absorbed through the skin are later excreted to some degree in exhaled air. Metabolites of some of these chemicals may also be exhaled. Examples of such chemicals include acetone, benzene, carbon tetrachloride, diethyl ether, ethanol, styrene, toluene, xylene, and many others. The concentrations of these chemicals in the exhaled air, however, has been found to decrease exponentially with time. Despite the fact that this is the method of preference for most workers, studies to date indicate that this measurement technique has some shortcomings. Although the body tends to integrate the external exposure, this monitoring technique reflects principally the most recent exposure level. The decay curves for many of these chemicals and times of exposure are frequently indistinguishable within 2 h after exposure, and it is impractical to instruct workers to collect alveolar air samples after work and return for analysis the following day.

Urinalysis

Urinalysis is probably the second most preferred method of biological monitoring among workers because it is not invasive. It usually relies on analysis for a metabolite in most organic chemicals. However, there have been great variations reported between authors for the translation of urinary levels of the index chemical to exposure levels of the workplace chemical. Despite the disagreement in translating urinary levels to body burden, there is no question that for many chemicals a relationship does exist. Problems could arise in urinalysis when the index chemical may come from diet, medication, or some other nonoccupational sources. For example, workers should not have taken Chloraseptic® (Richardson-Vicks Inc./Procter & Gamble), which contains sodium phenolate, prior to biological monitoring for benzene exposure.

Blood Analysis

Blood analysis is an invasive technique that carries some resultant risk to the worker. Most workers find it objectionable and would object strongly to phlebotomy or blood drawing. Generally, the order of worker acceptance for sampling for biological monitoring would be expected to be exhaled air, urine collection, and last of all, phlebotomy. there is no question that total blood levels of index chemicals reflect total dosage. For example, the correlations of inorganic lead exposure, biological effects, and blood concentrations for inorganic lead have been extensively reviewed.

Miscellaneous Analysis

Human breast milk analysis has not been used to monitor exposure to industrial chemicals even though milk could be considered as an excretory mechanism. Because of the small number of lactating women in the workplace, human breast milk analysis has very restricted utility for monitoring industrial chemical exposure.

Saliva, tears, perspiration, hair, nail, and adipose tissue, although not usually considered excretory fluids, do contain chemicals transferred from blood. To date, very little meaningful work has been reported on the analysis of these tissues as a measure of industrial exposure.

REFERENCES

Hall, S. K., Biological monitoring of solvent exposure, *Pollution Engineering*, 21(6):124–129, 1989.

Hall, S. K., Markiewicz, D. S., and Sherman, L. D., Biological monitoring of workers exposed to hazardous materials, *Proceedings of the Third Annual Presentation of the Hazardous Materials Management Conference*, Rosemont, IL, 1990.

Ho, M. H. and Dillon, H. K. (eds), *Biological Monitoring of Exposure to Chemicals*, John Wiley & Sons, Inc., New York, 1987.

Kneip, T. J. and Crable, J. V. (eds), *Methods for Biological Monitoring*, American Public Health Association, Washington, DC, 1988.

Lauwerys, R. R. and Hoet, P., *Industrial Chemical Exposure — Guidelines for Biological Monitoring*, 2nd edition, Lewis Publishers, Boca Raton, FL, 1993.

20

Chemical Exposure and Risk Communication

Stephen K. Hall and Cathleen M. Crawford

CONTENTS

INTRODUCTION

The term "hazard" is used to describe any activity or phenomenon that poses potential harm or other undesirable consequences to some people or things. The magnitude of a hazard is the amount of harm that may result, including the number of people or things exposed and the severity of consequence. The concept of "risk" further quantifies hazards by attaching the probability of

1-56670-239-9/97/$0.00+$.50
© 1997 by CRC Press, Inc.

realization at each level of potential harm. Thus, an area that experiences a severe earthquake once in 100 years faces the same hazard but only 1/10 the risk of a similar area that experiences an equally severe earthquake once in 10 years. Therefore, the concept of risk makes clear that hazards of the same magnitude do not always pose equal risks. "Risk assessment" is the characterization of potential adverse effects of exposures to hazards including estimates of risk and of uncertainties in measurements, analytical techniques, and interpretive models. Quantitative risk assessment characterizes the risk in numerical representations. The concept of health risk assessment has been reviewed in Chapter 16.

"Risk management" is the term generally used to refer to the characterization of the potential adverse effects of exposures to hazards, including estimates of risk and of uncertainties in measurements, analytical techniques, and interpretive models. In the past, the term risk communication has commonly been thought of as consisting only of one-way risk messages from experts to nonexperts. Today, however, "risk communication" is viewed as an interactive process of exchange of information and opinion among individuals, groups, and institutions. Often, it involves multiple messages about the nature of risk, or expressing concerns, opinions, or reactions to risk messages or to legal and institutional arrangements for risk management. A "risk message" is a written, verbal, or visual statement containing information about risk, and it may or may not include advice about risk reduction behavior.

In a democratic society, the government is required to inform the public. A series of federal laws, beginning with the Administrative Procedures Act of 1946, continuing with the Freedom of Information Act, the National Environmental Policy Act, and the Community Right-to-Know provisions of Title III of the Superfund Amendments and Reauthorization Act of 1986, recognizes the public's right to be informed about certain hazards and risks. Many of these laws have been reinforced by federal court decisions and presidential executive orders. These actions emphasize the government's responsibility to be accountable to the people. Thus, federal agencies are required to send messages to the public about the reasons for their decisions and to solicit comments from citizens. Regulators must explain why one course was chosen rather than another; the public has a right to see and challenge the basis for the decisions.

Frequently, only a few experts posses the best knowledge available to estimate accurately the extent of possible harm or the likelihood of its occurrence. To remain democratic, a government agency or industry must find ways to put specialized knowledge into the service of public choice. Because technological decisions have implications for public health and for political power, they are often highly contentious and emotional. Government and industrial officials are frustrated when their discussions are met by expressions of public mistrust and accusations of maleficence. Technical experts are frustrated when their explanations of available knowledge are met with apathy, disbelief, or anger. Nonexperts are frustrated by the inaccessibility of the knowledge they need to form their opinions, and by presentations of needed knowledge that are oversimplified, overly technical, or condescending in tone.

Negatively stated, not all risk communications are designed to inform or obtain understanding. The objective may be to maintain or foster ignorance; to hide or feign feelings, rather than to profess them; to lead astray rather than to guide; to give the next best advice, rather than the best; to obscure, to explain inadequately, to oversimplify, to slant, to popularize; or to tell only part of the truth, to mask it, or simply to lie.

MISCONCEPTIONS ABOUT RISK COMMUNICATION

Several important misconceptions need to be dispelled before the strategies in risk communication can be addressed. Many people—including scientists, decision makers, and members of the public—have unrealistic expectations about what can be accomplished by risk communication. For example, it is mistaken to expect improved risk communication to always reduce conflict and smooth risk management. Risk management decisions that benefit some citizens can harm others. In addition, people do not lead to consensus about controversial issues or to uniform personal behavior. However, even though good risk communication cannot always be expected to improve a situation, poor risk communication will nearly always make it worse.

It is also mistaken to think, as some do, that if people understood and used risk comparisons it would be easy for them to make decisions. Comparing risks can help people comprehend the unfamiliar magnitudes associated with risks, but risk comparison alone cannot establish levels of acceptable risk or ensure systematic minimization of risk. Factors other than the level of risk—such as the voluntariness of exposure to the hazard and the degree of dread associated with the consequences—must be considered in determining the acceptability of risk associated with a particular activity or phenomenon.

Some risk communication problems derive from mistaken beliefs about scientific research. Scientific information, for example, cannot be expected to resolve all important risk issues. All too often research that would answer the question has not been conducted, or the results are disputed. Although a great deal of research has been done on the dissemination and preparation of risk messages, there has been much less attention devoted to the risk communication process. In addition, even when valid scientific data are available, experts are unlikely to agree completely about the meaning of these data for risk management decision.

Other misconceptions involve stereotypes about the way recipients react to risk messages. It is mistaken, for example, to view journalists and the media always as significant, independent causes of problems in risk communication. Rather, the problem is often at the interface between science and journalism. Both sides need to better understand the pressures and constraints of the other instead of complaining about the sometimes disappointing results. Scientists and risk managers should recognize the importance of the part journalists play in identifying disputes and maintaining the flow of information during resolution of conflicts. Journalists need to understand how to frame the technical and social dimensions of risk issues. It is also important to recognize the differences between the broadcast and the print media and between the national and the regional local press corps.

Finally, even though most people prefer simplicity to complexity, it is incorrect to expect the public to want simple, cut-and-dried answers as to what to do in every case. The public is not homogeneous. People differ in the degree to which they exercise control over exposure to hazards or remediation of undesirable consequences, the importance they attach to various consequences, and their tendency to be risk adverse or risk seeking.

THE SCIENCE OF RISK ANALYSIS

The science one learned in school offers relatively tidy problems. The typical exercise in physics, for example, gives all the facts needed for its solution, and nothing but those facts. The difficulty of such problems for students comes in assembling those facts in a way that provides the right answer. The same assembly problem arises when analyzing the risks and benefits of a hazard. Scientists must discover how its pieces fit together. They must also figure out what the pieces are.

Understanding the science of risk analysis requires a series of essential steps. The first is to identify the scope of the problem under consideration, in the sense of identifying the set of factors that determine the magnitude of the risks and benefits produced by an activity or technology. The second step is to identify the set of widely accepted scientific facts that can be applied to the problem. The third step in understanding the science of risk analysis is to know how it depends on the educated intuitions of scientists, rather than on accepted hard facts.

Laypeople trying to follow a risk debate must understand how various groups of scientists have defined their pieces of the problem. Even the most accomplished of scientists are laypeople when it comes to any aspects of a risk debate outside the range of their trained expertise. The difficulties of determining the scope of a risk debate emerge quite clearly when one considers the situation of a reporter assigned to cover a risk story. The difficult part of getting most environmental stories is that no one person has the entire story to give. Such stories typically involve diverse kinds of expertise so that a thorough journalist might have to interview specialists in toxicology, epidemiology, economics, groundwater movement, meteorology, and emergency response, not to mention a variety of federal, state, and local officials concerned with public health, civil defense, education, and transportation. Even if a reporter consults with all the relevant experts, there is no assurance of complete understanding and coverage.

Some protocols that can be used in conducting a risk analysis are the causal model, the fault tree model, the flow diagram model, and a risk analysis checklist. The causal model of hazard creation is a way to organize the full set of factors leading to and from an environmental mishap, both when getting the story and when telling it. The example in Figure 20.1 is an automobile accident, traced from the need for transportation to the secondary consequence of the collision. Between each stage, there is some opportunity for an intervention to reduce the risk of an accident. By organizing information about the hazard in a chronological sequence, this scheme helps ensure that nothing is left out, such as the deep-seated causes of the mishap to the left and its long-range consequences to the right.

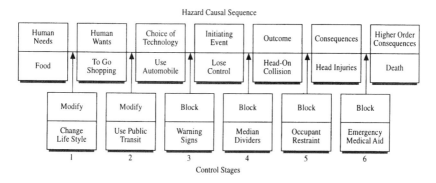

FIGURE 20.1 Causal chain of hazard evaluation. Reading from left to right, the top line indicates seven stages of hazard development, and the bottom line indicates six control stages.

A variant on this causal model is the fault tree, which lays out the sequence of events that must occur for a particular accident to happen. The example in Figure 20.2 indicates the possible ways that radioactivity could be released from deposited radioactive wastes after the closure of a repository. In effect, fault trees break open the right-handed parts of a causal model for detailed treatment. They can help a reporter to order the pieces of an accident story collected from different sources, see where an evolving incident, e.g., Three Mile Island, is heading, and find out what safety measures were or were not taken.

The flow diagram model is adapted from the engineering notion of a materials or energy flow diagram. The example in Figure 20.3 indicates the current options for the nuclear fuel cycle. If something is neither created nor destroyed in a process, then one should be able to account schematically for every bit of it. In environmental affairs, one wants to account for all toxic materials. It is important to know where each toxic agent comes from and where each goes.

A risk analysis checklist is a list of questions that can be asked in a risk analysis in order to clarify what problem has been addressed and how well it has been solved. Unfortunately, how one asks the question may in large part determine the content and apparent wisdom of the response. In principle, a risk analysis checklist does no more than organizing information from substantive disciplines in a way that allows overall estimates of risk to be computed. If it

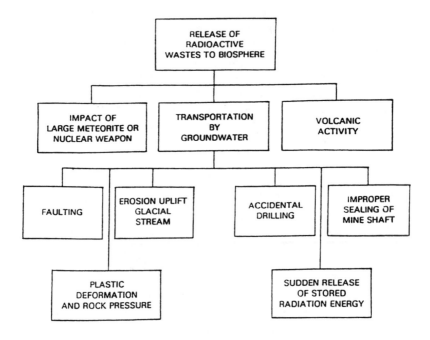

FIGURE 20.2   Fault tree indicating the possible ways that radioactivity could be released from deposited wastes after the closure of a repository.

is done correctly, it can facilitate citizen access by forcing all the facts out on the table. However, unless one can penetrate all its formalisms, a risk analysis checklist may mystify and obscure the facts rather than revealing them.

PSYCHOLOGY IN RISK COMMUNICATION

Whenever they read something in print, talk to their neighbors, or observe ominous activities at a local plant in order to understand the risks of a technology, people must rely on the same basic cognitive processes that they use to understand other events in their lives. The following is a number of generally supported statements about the behavior of people.

Most substantive decisions require people to deal with more nuances and details than they can readily handle at any one time. People have to juggle a multitude of facts and values when deciding, for example, to protest a landfill. To cope with this information overload, people simplify. Rather than attempting to think their way through to comprehensive, analytical solutions to decision making, people tend to rely on habit, tradition, the advice of neighbors, the news media, and on general rules of thumb. Rather than considering the extent to which human behavior varies from situation to situation, people describe other people in terms of all-encompassing personality traits such as being honest, happy, or risk seeking. Rather than think precisely about the

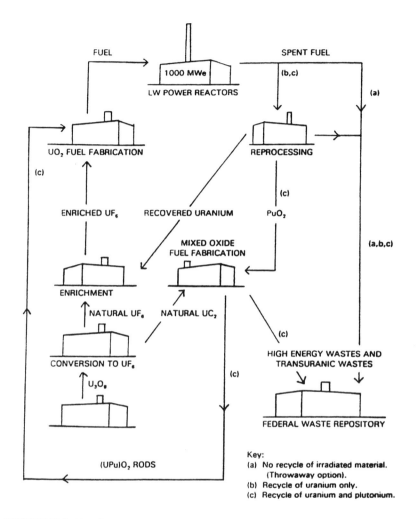

FIGURE 20.3 Materials and energy flow diagram: options for the nuclear fuel cycle.

probability of future events, people rely on vague quantifiers, such as "likely" or "not worth worrying about"—terms that are also used differently by different people and by the same individual in different contexts. People's desire for simplicity can be observed when they press risk managers to categorize technologies as "safe" or "unsafe," rather than treating safety as a continuous variable. Although such simplifications help people cope with life's complexities, they can also obscure the fact that most risk decisions involve gambling with people's health, safety, and economic well-being in arenas with diverse actors and shifting alliances.

Fortunately, given their need to simplify, people are quite good at observing those events that come to their attention. As a result, if the appropriate facts reach people in a responsible and comprehensible form before their minds are made up, there is a good chance that their first impression will be the correct one. For example, most people's primary sources of information about risks are what they see in the news media and observe in their everyday lives. Consequently, people's estimates of the principal causes of death are strongly related to the number of people they know who have suffered those misfortunes and the amount of media coverage devoted to them. Unfortunately, what most people see are the outward manifestations of the risk management process, such as hearings before regulatory bodies or statements made by scientists to the news media. In many cases, these outward signs are not very reassuring. Often, they reveal acrimonious disputes between supposedly reputable experts, accusations that scientific findings have been distorted to suit their sponsors, and confident assertions that are disproved by subsequent research.

Not all problems with information about risks are as readily observable as blatant lies. Often, the information that reaches the people is true, but only partially so. Detecting such systematic omissions is quite difficult for laypeople. As a result of this insensitivity to omission, people's risk perceptions can be manipulated in the short run by selective presentation. Not only will people not know what they have not been told, but they will not even notice how much has been left out. What happens in the long run depends on whether the unmentioned risks are revealed by experience, or by other sources of information. When deliberate omissions are detected, the responsible party is likely to lose all credibility. Once a shadow of doubt has been cast, it is hard to erase.

The nuclear disaster at Chernobyl is a textbook example. In the first months after the explosion, it was announced that more than 100,000 people were evacuated from an 18-mile radius around the plant. Then the officials admitted to an appalling underestimation of the danger of radiation in a much broader region and 200,000 more people had to be uprooted. Even the government newspaper *Izvesta* admitted that as many as 3 million people are living on irradiated land. Health statistics showing a sharp increase in cancer and other radiation-related diseases and abnormalities are emerging after several years of bureaucratic obfuscation. In the politically charged atmosphere of the *perestroika* era, people are ready to believe the worst.

Despite their frequent intensity, risk debates are typically conducted at a distance. The debating parties operate within self-contained communities and talk principally to themselves. Opponents are seen primarily through their writing or their posturing at public events. Thus, the public's insensitivity to the importance of how risk issues are presented exposes itself to manipulation. For example, a risk might seem much worse when described in relative terms than in absolute terms (e.g., doubling their risk versus increasing that risk from one in a million to one in a half-million). Although both representations of the risk might be honest, their impacts would be quite different.

Finally, the public is extraordinarily adept at maintaining faith in its current beliefs unless confronted with concentrated and overwhelming evidence to the contrary. One psychological process that helps people maintain their current beliefs is feeling little need to look actively for contrary evidence. Why look, if one does not expect that evidence to be very substantial or persuasive? A second contributing thought process is the tendency to exploit the uncertainty surrounding apparently contradictory information in order to interpret it as being consistent with existing beliefs. A third thought process can be found in the public's reluctance to recognize when information is ambiguous.

The above statements reduce both complex people and intricate research literatures to necessarily oversimplified summaries. Neither the public nor the literature can be categorized without their appropriate context. Ideally, one would have polished studies of how specific people respond to specific risks. This should be the standard for designing and evaluating risk communication programs.

CONTENT OF RISK MESSAGES

Risk messages should closely reflect the perspectives, technical capacity, and concerns of the target audience. A risk message should emphasize information relevant to any practical actions that individuals can take, be couched in clear and plain language, respect the audience and its concerns, and seek to inform the recipient of the message. One of the difficult issues in risk communication in a democratic society like ours is the extent to which public officials should attempt to influence the recipients' beliefs and risk-reducing actions.

Uncertainty is a fact in risk assessment. It is also present in risk management options. The way that risk messages treat this uncertainty can have a major influence on the effectiveness and credibility of a communication effort. Thus, risk messages and supporting materials should not minimize the existence of uncertainty. Data gaps and areas of significant disagreement among experts should be disclosed. Some indication of the level of confidence of estimates and the significance of scientific uncertainty should be conveyed.

One factor that inhibits public understanding of risk messages is that people often cannot easily relate risk probabilities to their everyday experience. Risk comparisons can be helpful but they should be presented with caution. Risk comparisons must be seen as only one of several inputs to risk decisions, not as determinants of decisions. There are proven pitfalls when risks of diverse character are compared, especially when the intent of the comparison can be seen as that of minimizing a risk. More useful are comparisons of risks that help convey the magnitude of a particular risk estimate that occur in the same decision context (e.g., risks from landfilling household waste versus hazardous waste). Multiple comparisons may help avoid some of the pitfalls.

If the information in a risk message is incomplete, the recipients may be unable to make well-informed decisions. Thus, a complete risk message should contain information on (1) the nature of the risk; (2) the nature of the benefits that might be affected if risk were reduced; (3) the available alternatives; (4) the uncertainty in knowledge about risks and benefits; and (5) the management issues. In addition, there are major advantages in putting the information into written form as an adjunct to the risk message.

STRATEGIES IN RISK COMMUNICATION

The technical and policy issues involved in making risk management decisions are complex enough in themselves. Dealing with public perceptions of risks creates an additional level of complexity for risk managers. Following are some of the more common basic strategies for dealing with risk controversies.

Dealing with Facts

The assumption underlying this strategy is that if the public only knew as much as the experts, they would respond to the hazards in the same way. Undertaken insensitively and insensibly, this strategy can result in an incomprehensible deluge of technical details, telling the public more than it needs to know about specific risk research results, and much less than it needs to know about the quality of the research.

Persuading the Public

The premise here is that the public needs persuasion, rather than education. It often follows the failure of an information campaign to win public acceptance for a technology. Undertaken heavy-handedly, this approach may amount to little more than repeating more loudly messages that the public has already rejected. Here, as elsewhere, obvious attempts at manipulation can breed resentment.

Relating to the Past

The underlying assumption here is that the public will accept in the future the kinds of risks that it has accepted in the past. If this is true, then what the public will accept can be determined simply by examining statistics showing the risk-benefit trade-offs involved in existing technologies. This assumption ignores the fact that the public may be unhappy with how risks have been managed in the past. As a result, giving them more of the same means enshrining past inequities in future decisions.

Involving the Local Communities

This approach assumes that people will be flexible and realistic about trade-offs when they have responsibility for the big picture. Such an approach can flounder when the local community lacks real decision-making authority or the technical ability to understand its alternatives. It may also flounder when those alternatives accept perceived past inequities. Ensuring the informed consent of the public for the risks to which they are exposed is a laudable goal. However, its achievement requires that the public have tolerable choices, adequate information, and the ability to identify which course of action is in their own best interests.

Explaining Regulatory Philosophy

The assumption here is that the public does not want facts, but instead the assurance that they are being protected. That is, whatever the risks may be, they are in line with government policy. For example, the Nuclear Regulatory Commission's "safety goals for nuclear power" describes how risky it will allow the technology to be. The policy is stated in terms of levels of acceptable risk, as though the public is too unsophisticated to understand. It should be noted that overly simplified statements provide little guidance for real statements. They simply deny the complexity of the risk-benefit decisions that needed to be made. If perceived as hollow, then the public is not assured.

Designating a Spokesperson

The assumption underlying this strategy is that what the public needs in order to understand risk issues is a coherent story from a single credible source. An example is the assumption of center stage by the president of Union Carbide after the chemical disaster in Bhopal, India. This strategy can reduce the confusion created by incomplete and conflicting messages. This strategy only works well if the spokesperson has good communication skills or is sensitive to listeners' information needs; the person must have both substance and style. Oversimplifications, misrepresentations, and unacceptable policies are just that, even if they come from a charming person.

MANAGEMENT OF RISK COMMUNICATION

Much concern about risk communication has been focused on questions of message content. Failures have frequently been attributed to the inability of the audience to comprehend complex technical issues and to the tendency of risk messages to be badly written. This view has led many risk managers to seek solutions in the design of better risk messages themselves. However, the process by which risk management decisions are made and explained has gained much attention. Four process objectives have been identified as the key elements in improving risk communication: goal setting, openness, accuracy, and competence.

Setting Realistic Goals

Risk communication activities ought to be matters of conscious design. Realistic goals should be established that explicitly accommodate the legal mandates bounding the process, and the roles of the potential recipients of the organization's risk messages. Explicit consideration of such factors encourages realistic expectations, clarification of motives and objectives—both within the source organization and among outside groups and individuals—and evaluation of performance.

Safeguarding Openness

In many cases, risk communication efforts have floundered because public trust and credibility were lacking. Risk communication should be a two-way street. Organizations that communicate about risks should ensure openness as well as an effective dialogue with the public. This two-way process should be characterized by a spirit of open exchange in a common undertaking, rather than a series of "canned" briefings restricted to technical issues. In addition, there should be an early and sustained interchange that also includes the media and other message intermediaries. Openness does not imply empowerment to determine the organization's risk management decisions. To avoid misunderstanding, the limits of participation should be made clear from the outset.

Safeguarding Accuracy

For many risk messages, credibility depends on the audience's belief that the message is reasonably accurate. There is always broad skepticism about an organization shading the truth to suit its ends. To help ensure that risk messages are not distorted and do not appear to be distorted, those who manage the generation of risk assessments and risk messages should hold the preparers of messages accountable for detecting and reducing distortion. When feasible, recognized independent experts should review the risk assessment and message. Organizations should be prepared to release for comment—a "white paper" on the risk assessment as well as risk reduction assessment.

Fostering Competence

Risk communication has only recently come into focus as a concept and in many organizations it is still managed under other functions such as public affairs. More attention should be paid to risk communication as a distinct undertaking. Successful efforts in risk communication require a blend of technical and communications proficiency. Organizations should take steps to ensure that the preparation of risk messages to be a deliberate, specialized undertaking, taking care that in the process they do not sacrifice scientific quality. Where there is a foreseeable potential for emergency, advance plans for communication should be drafted.

SUCCESSFUL RISK COMMUNICATION PROGRAMS

Performing an evaluation of risk communication programs requires a clear, operable definition of the consequences to be desired. With medical treatment, for example, identifying the consequences is usually a straightforward process. Accompanying the treatment are various possible health effects, some good and some bad. Although medical personnel and their patients are likely to agree about which outcomes are good and which are bad, they need not agree about how good and how bad the outcomes are.

In evaluating risk communication programs, similar issues arise. Potential consequences must still be identified. However, the issues are less clearly defined. There are also the good and bad health effects but they may be harder to be defined. Difficulties in observing the effects of ultimate interest may divert attention to more observable effects closer to the problem. One possibility in evaluation is assessing comprehension of the message. If people have not understood the message, then an appropriate response would be unlikely. The simplest test of comprehension might be remembering the facts of a message. Those who pass the test would still have to be tested for whether they are able to use the retained facts in their decision making. Those who fail the test would still have to be tested for whether they have heard the message, but chose to reject it. Rejection might mean distrusting the source's competence or its motives. Thus, a risk communication program is considered successful if it raises the level of understanding of relevant issues or actions and satisfies those involved that they are adequately informed within the limits of available knowledge.

It is important to make several points about the definition of successful risk communication programs. First, success is defined in terms of the information available to the decision makers, rather than in terms of the quality of the decisions that follow. In other words, successful risk communication does not always lead to better decisions, because risk communication is only part of risk management. Second, successful risk communication need not result in consensus about controversial issues or in uniform personal behavior. Third, according to the definition of success, the recipients of a risk message must be able to achieve as complete an understanding of the information as they desire. A risk communication program is not successful unless the recipients achieve a sufficient understanding. Thus, the success of a risk communication program should not be judged by the level of knowledge reflected in particular messages accessible to the decision maker, but by the level of knowledge on which decision-makers act.

REFERENCES

Hall, S. K. and Crawford, C. M., Risk analysis and risk communication, *Pollution Engineering*, 24(19):78–83, 1992.

Morgan, M. G., Risk analysis and management, *Scientific American*, 269:32–45, 1989.

National Research Council, *Improving Risk Communication*, National Academy Press, Washington, DC, 1989.

National Research Council, *Issues in Risk Assessment in the Federal Government*, National Academy Press, Washington, DC, 1993.

Environmental Protection Agency, General Quantitative Risk Assessment Guidance for Non-cancer Health Effects, ECAP-CIN-538M, 1989.

Index

A

Absorption. *see also specific toxin*
  mechanisms of, 10
  routes of
    dermal, 11
    description, 11
    gastrointestinal, 12–13
    respiratory, 13–14
Acetaminophen, nephrotoxicity of, 129
2-Acetylaminofluorene, metabolism of, 229, 232
Acetylcholinesterase, 146
Acrylonitrile, occupational exposure, 59
Active transport, 10
Additive reaction, 8
ADH. *see* Antidiuretic hormone
Aerosol, 93
AHH. *see* Aryl hydrocarbon hydroxylase
Alcohols
  biological monitoring, 248
  description of, 45
  industrial use of, 45
  metabolism of, 45
  toxic effects of, 45–46
Aldehydes, biological monitoring, 248
Alicyclic hydrocarbons
  biological monitoring, 248–249
  description of, 42
  oxidation of, 248
  toxic effects of, 42–43
Aliphatic amines
  absorption of, 50
  biological monitoring, 252
  description of, 50
  industrial uses of, 50
  oxidation of, 249, 252
Aliphatic hydrocarbons
  description of, 41
  toxic effects of, 41–42
Alkanes, 41–42
Alkenes, 42

Alkylating agents
  biological properties of, 223
  reproductive toxicant effects of, 172
Allergic reaction, 8
Aluminum
  absorption of, 24, 239
  biological monitoring profile, 239–240
  chelation therapy, 24
  neurotoxicity associated with, 24
  occupational exposure, 93
  respiratory system effects, 24, 93
4-Aminobiphenyl
  cancer risk associated with, 54
  occupational exposure, 54
Amitrole, 59
Ammonia
  description of, 82
  occupational exposure, 82
  toxic effects, 82–83
Amphotericin B, nephrotoxicity of, 130
Analgesics, 129
Androgens, 180–181
Aniline
  absorption of, 51
  description of, 51
  industrial uses of, 51
Animal proteins
  characteristics of, 86
  occupational exposure, 86
  toxic effects, 86
Antagonistic reaction, 8
Antibiotics
  immunosuppression, 207
  nephrotoxicity of, 129–130
Antidiuretic hormone, 124
Antimony
  biological monitoring profile, 240
  chronic poisoning, 25
  common compounds, 25
  concentration variations, 240
  oxidation states, 25
  toxic effects of, 25

I

Idiosyncratic reaction, 8
Immediate reaction, 8
Immune system
  cellular components
    lymphocytes, 202–203
    macrophages, 202
    neutrophils, 203
  chemical mediators
    complement, 203
    immunoglobulins, 203
    lymphokines, 203
  description of, 202
  hypersensitivity, 204
  laboratory assessment of
    cellular immunity, 204
    complement assessment, 204
    humoral immunity, 204
    phagocytosis, 203
  responses to toxins, 204
  toxic responses
    allergic reactions, chemicals that cause
      cosmetics and enzymes, 205
      detergents, 205
      formaldehyde, 205
      metals, 205
      resins, 205
    autoimmunity drugs
      antibiotics, 207
      hydralazine, 207
      metals, 207
      methyldopa, 207
      phenytoin, 207
    environmental pollutants
      aromatic hydrocarbons, 206
      benzene, 206
      metals, 206
      ozone, 206
      pesticides, 206
      urethane, 206
    immunosuppression
      alcohol, 205
      cannabis, 206
      cocaine, 205
      cyclosporine A, 205
      cytotoxic drugs, 205
      heroin, 206
      nitrites, 206
      steroids, 205
    neoplastic changes
      aromatic hydrocarbons, 207
      asbestos, 207

    benzidine, 207
    vinyl chloride, 207
Immunoglobulins, 203
Insecticides
  cholinesterase-inhibiting
    mechanism of action, 148
    symptoms of, 148–150
  organochlorine
    absorption routes, 150–151
    DDT. *see* Chlorophenothane
    definition of, 150
    EPA ban of, 150
    neurotoxicity of
      mechanism of action, 150
      symptoms of, 150–151
    types of, 149
  organophosphorus
    neurotoxicity of
      mechanism of action, 148
      symptoms of, 148–150
    types of, 149
Intracellular water, 14
Intrastitial water, 14
Ionization
  effect on absorption of toxicants, 9
  formulas to determine, 9
Iron
  absorption of, 241
  chelation therapy, 31
  chronic exposure to, 31
  excretion of, 241
  function of, 31
  occupational exposure, 95
  pulmonary exposure to, 31
  toxic exposure. *see* Siderosis
Iron ore and oxides. *see* Hematite
Irreversible reaction, description of, 8
Ischemic necrosis, 193
Isopropyl alcohol
  cancer risk associated with, 57
  commercial uses of, 57
  description of, 45
  occupational exposure, 57
Itai-itai disease
  description of, 28
  symptoms of, 28
Ito cells, 113

J

Jaundice
  cholestasis and, 115
  definition of, 115–116

K